Problems – Ideas - Solutions

Proof of the Collatz conjecture

Georgiy Tyshko

Problems – Ideas - Solutions

How Do We, People on Earth, Want to Live Tomorrow?

What kind of life would we like to see around us?
Making the life around us better, more comfortable,
enjoyable, and effective in every way depends on us, on each
and every one of us.

Let's build our future ourselves

Book Six – Proof of the Collatz conjecture

1

Preface   9

1. Wording   12

2. The Reduced sequence of Collatz   12

3. Modified reduced sequence of Collatz   13

4. Representation of collatz and the types of collatz   15

5. Collatz Tree   15

6. The idea of the proof   18

7. Example of a fragment (subtree) of the Collatz tree   18

8. Vertical and Horizontal numbers   20

9. Branches and Trunks of the Collatz Tree   21

10. Full trunks and full branches   22

11. Direct and Reverse Vertical and Horizontal
Transformations of Collatz i   22

12. Direct and Reverse horizontal sequence of collatz is   24

    12.1 "n-Similarity" direct horizontal sequences of
    Collatz   25

    12.2 "n-Similarity" reverse horizontal sequences of
    Collatz   27

13. Rule $(3/4)^n$ and rule $(2/3)^n$ for horizontal
sequences with lengths greater than n.   29

    13.1 Rule $(3/4)^n$ for sequences whose lengths are
    greater than n in a set of direct horizontal sequences.   30

        13.1.1 the Rule $(3/4)^1$ for sequences with lengths
        that are greater than 1 in the set of direct horizontal
        sequences.   30

        13.1.2 Rule $(3/4)^2$ for sequences with lengths that

are greater than 2 in the set of direct horizontal sequences. 34

13.1.3 Rule (3/4)^n for sequences whose lengths are greater than n in a set of direct horizontal sequences. 47

13.1.3.1 Direct horizontal sequences in which the members n, n>1 are natural integers of type 5 47

13.1.3.2 Direct horizontal sequences in which the members n, n>1 are natural integers of type 1 64

13.1.3.3 Direct horizontal sequences in which the members n, n>1 are natural integers of type 7 82

13.1.3.4 Direct horizontal sequences in which the members n, n>1 are natural integers of type 11 88

13.1.3.5 Consequence from sections 13.1.3.1 to 13.1.3.4 95

13.2 Rule (2/3)^n for sequences whose lengths are greater than n in a set of reverse horizontal sequences.96

13.2.1 the Rule (2/3)^1 for sequences with lengths that are greater than 1 in the set of reverse horizontal sequences. 96

13.2.2 Rule (3/4)^2 for sequences with lengths that are greater than 2 in the set of reverse horizontal sequences. 100

13.2.3 Rule (2/3)^n for sequences whose lengths are greater than n in a set of reverse horizontal sequences. 108

13.2.3.1 Reverse horizontal sequences in which the members n, n>1 are natural integers of type 3 108

13.2.3.2 Reverse horizontal sequences in which the members n, n>1 are natural integers of type 5   108

13.2.3.3 Reverse horizontal sequences in which the members n, n>1 are natural integers of type 7   124

13.2.3.4 Reverse horizontal sequences in which the members n, n>1 are natural integers of type 9   140

13.2.3.5 Reverse horizontal sequences in which the members n, n>1 are natural integers of type 11   140

13.2.3.6 Reverse horizontal sequences in which the members n, n>1 are natural integers of type 1   156

13.2.3.7 Consequence from sections 13.2.3.1 to 13.2.3.6   171

14. The absence of "loops" in the horizontal sequences   **172**

14.1 The absence of "loops" in the direct horizontal sequences   172

14.1.1 Direct horizontal sequences that begin in natural integers of type 3 and 9   173

14.1.2 Direct horizontal sequences that begin in vertical natural integers   173

14.1.3 Direct horizontal sequences that end in vertical natural integers   174

14.1.4 Direct horizontal sequences that end in other natural integers   174

14.2 The absence of "loops" in the reverse horizontal sequences   176

15. All horizontal sequences are finite   **177**

15.1 All direct horizontal sequences are finite   177

15.1.1 The first proof of finiteness of direct horizontal sequences   177

15.1.2 The second proof of finiteness of direct horizontal sequences. Reverse Structure of the Collatz and Reverse Fibonacci Structure of the Collatz   177

15.2 All Reverse horizontal sequences are finite   197

15.2.1 The First proof of the finiteness of the reverse horizontal sequences   197

15.2.2 The second proof of finiteness of reverse horizontal sequences. Direct Structure of the Collatz197

15.3 complete horizontal and complete vertical sequences   205

16. The Uniqueness of The Canonical Tree of Collatz205

17. Afterword   209

Preface

In memory of my friend Bob Golubev.

I express my deepest gratitude to Elic Yavor, Evgeny Shakhnovich, Mikhail Gromov, Jon Links and Vera Dmitriyeva for the great financial assistance and moral support.

This book is at first glance a proof of the well-known conjecture of Lothar Collatz on the Syracuse sequence. However, in fact, this book is about finding consistency and regularity in the world around us.
Without any doubt, there will be many criticisms about the inconclusiveness of the proof, the presence of errors, the presence of inaccuracies, the presence of unnecessary minor details, inappropriate mathematical presentation, etc
However, both computing projects to search for a counterexample will be stopped because of the obvious allegiance to the Conjecture of Collatz is after the project management will familiarize themselves with the material of the book
Moreover, there will be new correct proofs of the validity of the Collatz Conjecture, which are quite possibly shorter and more correctly stated mathematically
However, all these new proofs will be very important to use the tool of Modified Reduced Sequences of Collatz, the tool of Canonical Tree of Collatz, the tool of Branches and Trunks of Collatz Tree, the tool of Vertical and Horizontal Sequences, the tool of Direct and Reverse structures of Collatz and other tools outlined in the book
Probably the tool of Vertical and Horizontal numbers as well as the tool of the types of the Collatz will be used in

attempts to solve other unsolved problems of number theory.

The most important value of the material presented in the book is precisely in the detection of the tool types of the Collatz and it is in the detection of the tool Horizontal and Vertical numbers.

The material of the book shows how, as a result of minor transformations, the chaos of the "hailstone numbers" behavior turns into a coherent and regular picture.

The study of the Canonical Tree of the Collatz and the Direct And Reverse structures of the Collatz in itself is a very interesting direction in the development of number theory in particular and in the harmony and regularity of the world in General.

The material of the book is clear and accessible to any school child from 11-12 years.

The material of the book can be a source of a huge number of tasks for programming Olympiads.

On the example of the problem of correctness of The Collatz conjecture about the Syracuse sequence, I want to draw attention to the following circumstance and propose a new paradigm for solving any problems (not only in mathematics, but also in engineering and in General in all human activities)

The fact is that until now, mankind has meant two intellects, namely the usual human intelligence and artificial intelligence of computers.

While there is actually many times more powerful intellect than both of the above mentioned intellects, namely there is a Collective intellect.

A proof of the validity of the Collatz hypothesis, devoid of any drawbacks, could have been obtained within a few weeks if the Collective Intellect had set itself the task of building such a proof

Therefore, another goal of this book is the author's desire to create and develop a paradigm of Collective Intellect.

Currently, the most important prerequisite for the creation and development of the paradigm of Collective Intellect has appeared.

It's about the rapidly growing ability of people to communicate in virtual reality.

In the very near future I want to create appropriate virtual platforms based on Second Life, Sansar, Decentraland or any other virtual reality.

However, while such virtual platforms are not created, I suggest everyone to leave their questions and comments in the blog howwewanttolive.livejournal.com

In this blog I will answer any questions and comments on the material of this book as well as on all other topics covered in this blog.

I am not a professional mathematician and currently I live on small pension in abject poverty. If someone like my ideas and solutions and such a person is ready to support me financially then it can be done through my account georgiytyshko@yahoo.com in PayPal.com

I am also grateful for the help and support of Alexey Muravitsky, Vsevolod Kaspler and Vladimir Rozhin

1. Wording

The conjecture of collatz is (conjecture 3n+1 conjecture 3x+1, collatz is the problem, the problem 3n+1 problem 3x+1, the Syracuse problem) — one of the unsolved problems of mathematics, named after the German mathematician Lothar collatz is the who suggested it in 1937. To explain the hypothesis, consider the following sequence of numbers, called the Syracuse sequence. Take any positive integer n. If it is even, then divide it by 2, and if it is odd, then multiply by 3 and add 1 (get 3n + 1). Over the resulting number perform the same actions, and so on. For example, for the number 3 we get:

3-odd, $3 \times 3 + 1 = 10$
10-even, $10 : 2 = 5$
5-odd, $5 \times 3 + 1 = 16$
16-even, $16 : 2 = 8$
8-even, $8 : 2 = 4$
4-even, $4 : 2 = 2$
2-even, $2 : 2 = 1$
1-odd, $1 \times 3 + 1 = 4$

Obviously, starting with 1, the numbers 1, 4, 2 begin to repeat cyclically.

2. The Reduced sequence of Collatz

Next, under the sequence of Collatz we understand the Syracuse sequence

Put in correspondence of each sequence of Collatz a new sequence of numbers obtained by throwing out the original sequence of all even numbers

We will call such a sequence of Reduced sequence of

Collatz

For example, sequences

27, 82, 41, 124, 62, 31, 94, 47, 142, 71, 214, 107, 322, 161,
484, 242, 121, 364, 182, 91, 274, 137, 412, 206, 103, 310,
155, 466, 233, 700, 350, 175, 526, 263, 790, 395, 1186,
593, 1780, 890, 445, 1336, 668, 334, 167, 502, 251, 754,
377, 1132, 566, 283, 850, 425, 1276, 638, 319, 958, 479,
1438, 719, 2158, 1079, 3238, 1619, 4858, 2429, 7288,
3644, 1822, 911, 2734, 1367, 4102, 2051, 6154, 3077,
9232, 4616, 2308, 1154, 577, 1732, 866, 433, 1300, 650,
325, 976, 488, 244, 122, 61, 184, 92, 46, 23, 70, 35, 106,
53, 160, 80, 40, 20, 10, 5, 16, 8, 4, 2, 1, ...

will match the reduced sequence

27, 41, 31, 47, 71, 107, 161, 121, 91, 137, 103, 155, 233,
175, 263, 395, 593, 445, 167, 251, 377, 283, 425, 319, 479,
719, 1079, 1619, 2429, 911, 1367, 2051, 3077, 577, 433,
325, 61, 23, 35, 53, 5, 1, ...

It is obvious that if the initial sequence "falls into the
number 1", then the reduced sequence also "falls into the
number 1" and vice versa

3. Modified reduced sequence of Collatz

Let's put a new sequence of numbers in accordance with
each reduced sequence of Collatz, obtained by including
new numbers in the original sequence

We will call such a sequence a Modified reduced sequence
of Collatz

12

New numbers will be included according to the following rule

If the number in the original sequence is a number of type 4*p+1 where p is odd, then the next number in the resulting sequence insert the number p, if the number is not, then the next number in the sequence leave the number that followed the number, if the inserted number p is also a number of type 4*q+1 where q is odd, then the number q is inserted after the number p and so on

For example, a reduced sequence

27, 41, 31, 47, 71, 107, 161, 121, 91, 137, 103 , 155, 233, 175, 263, 395, 593, 445 = 111*4+1, 167, 251, 377, 283, 425, 319, 479, 719, 1079, 1619, 2429 = 607*4+1, 911, 1367, 2051, 3077 = 769*4+1, 577, 433, 325 = 81*4+1, 61 = 15*4+1, 23, 35, 53 = 13*4+1, 5, 1, ...

the modified reduced sequence will match

27 , 41, 31, 47, 71, 107, 161, 121, 91, 137, 103, 155, 233, 175, 263, 395, 593, 445, 111, 167, 251, 377, 283, 425, 319, 479, 719, 1079, 1619, 2429, 607, 911, 1367, 2051, 3077, 769, 577, 433, 325, 81, 61, 15, 23, 35, 53, 13, 5, 1, ...

If the original reduced sequence "falls into the number 1", then the modified reduced sequence also "falls into the number 1" and Vice versa
Because $(3*(4*p+1)+1)/4 = 3*p+1$ and hence the images of the numbers 4*p+1 and p when displayed $(3*n+1)/2$ coincide

4. Representation of collatz and the types of collatz

Any natural odd number A can be represented as

$12*p+q$, where $p = 0,1, ...$ and $q = 1,3,5,7,9,11$

Let's call this representation of a number a the Representation of a Collatz

Let's call it the type of a natural odd number A, the number q in such representation

5. Collatz Tree

We call a Collatz tree generated by some natural odd number, a directed graph constructed as follows

Let p be the original natural odd number, then from the corresponding vertex always goes up the edge (oriented by the down arrow) to the vertex, which corresponds to the number $4*p+1$,

further, if p is a number of type 1 ($p=12*q+1$, where $q != 0$) , then from the corresponding vertex always goes to right the edge (oriented left arrow) to the vertex, which corresponds to the number $(4*(12*q+1) – 1)/3 = 16*q + 1$,

further, if p is a number of type 1 ($p=12*q+1$) and is a number representable as $4*t+1$ (where $t = 6*k+3$, $k=0,1,..$), then from the corresponding vertex always goes down the edge (oriented down arrow) to the vertex, which corresponds to the number t, further, if p is a number of type 1 (p=12*q+1, where q != 0) and is not a number representable as 4*t+1 (where t = 6*k+3) , then from the corresponding vertex always goes to the left the edge (oriented left arrow) to the vertex, which corresponds to the number $(3*(12 * q+1) +1)/4 = 9*q + 1$,

further, if p is a number of type 3 (p=12*q+3), then from the corresponding vertex always goes to the left the edge (oriented by left arrow) to the vertex, which corresponds to the number $(3*(12*q+3) +1)/2 = 18*q + 5$,

further, if p is a number of type 5 (p=12*q+5), then from the corresponding vertex always goes to the right the edge (oriented by the left arrow) to the vertex, which corresponds to the number $(2*(12*q+5) - 1)/3=8*q+3$,

further, if p is a number of type 5 (p=12*q+5) and is a number representable as 4*t+1 (where t = 6*k+1, k=0,1,..), then from the corresponding vertex always goes down the edge (oriented down arrow) to the vertex, which corresponds to the number t,

further, if p is a number of type 5 (p=12*q+5) and is not a number representable as 4*t+1 (where t = 6*k+1), then from the corresponding vertex always goes to the left the edge (oriented left arrow) to the vertex, which corresponds to the number $(3*(12*q+5) +1)/4 = 9*q + 4$,

further, if p is a number of type 7 (p=12*q+7), then from the corresponding vertex always goes to the right the edge (oriented by the left arrow) to the vertex, which corresponds to the number $(4*(12*q+7) - 1)/3=16*q+9$, and to the left the edge (oriented left arrow) to the vertex, which corresponds to the number $(3*(12*q+7)+1)/2 = 18*q+11$,

15

further, if p is a number of type 9 (p=12*q+9) and is a number representable as 4*t+1 (where t = 6*k+5, k=0.1,..), then from the corresponding vertex always goes down the edge (oriented down arrow) to the vertex, which corresponds to the number t,

further, if p is a number of type 9 (p=12*q+9) and is not a number represented as 4*t+1 (where t = 6*k+5), then from the corresponding vertex always goes to the left the edge (oriented by the left arrow) to the vertex,which corresponds to the number (3*(12*q+9)+1)/4 = 9*q+7,

further, if p is a number of type 11 (p=12*q+11), then from the corresponding vertex always goes to the right the edge (oriented by the left arrow) to the vertex, which corresponds to the number (2*(12*q+11) - 1)/3=8*q+7, and to the left the edge (oriented by the left arrow) to the vertex, which corresponds to the number (3*(12*q+11)+1)/2 = 18*q+17

It follows from the definition of the Collatz tree that the path is started at any vertex and is continued in the direction of the arrows to other vertex corresponds to a modified reduced sequence of the Collatz which started in a number that corresponds to the first vertex of the path and finished in a number that corresponds to the last vertex of the path

The Collatz tree generated from the number 1 will be called Canonical

6. The idea of the proof

The idea of the proof is to show that ALL the trees of the

Collatz which are generated by natural odd numbers are the same canonical tree of the Collatz generated by the number 1

In other words, the idea of the proof is to show that all natural odd numbers are present in the canonical tree of the Collatz, while the modified reduced sequence of the Collatz is started at any vertex of the canonical tree of the Collatz will ALWAYS fall in 1

7. Example of a fragment (subtree) of the Collatz tree

Everywhere further instead of numbers their types are shown

|
5-11- 7- 1- 5-11- 3
|
1- 5- 3
|
9
|
5- 3
|
1- 1- 1- 5- 7- 5- 3
|
9
|
5- 7- 5- 3
|
1- 9
|
9

|
5-11- 3
|
1- 5- 7- 1- 1- 5- 7- 1- 1- 1- 9
|
9
|
5- 3
|
1- 1- 9
|
9
|
5- 7- 9
|
1- 9
|
9
|
5-11-11- 3
|
1- 5-11- 7- 9
|
9
|
5- 3
|
1- 1- 5-11-11- 7- 9
|
9
|
5- 7- 1- 5- 3
|
1- 9
|
9

```
|
5-11- 7- 9
|
1- 5- 3
|
9
|
5- 3
|
1
```

A fragment of the canonical tree of a Collatz generated by the number 1

8. Vertical and Horizontal numbers

The natural odd number p is called Vertical if it can be represented as $4*q+1$ (where $q = 2*k+1$, $k=0,1,..$)
The number 1 is called the Root number
The rest of the natural odd numbers we will call Horizontal.

For any nonnegative p and any nonnegative odd integer q

$12*p+3 != 4 * q +1$, $12*p+7 != 4 * q +1$, $12*p+11 != 4*q +1$

hence all numbers of type 3, 7 and 11 are the horizontal numbers.

For any nonnegative even integer p and any odd non-negative q

$12*p+1 != 4 * q +1$, $12*p+9 != 4*q +1$

therefore, all numbers of type 1 and 9, if p is even, are horizontal numbers.

Accordingly, all numbers of type 1 and 9, if p is odd, are vertical numbers.

For any nonnegative p and any odd integer is odd non-negative q

12*p+5 != 4*q +1

therefore, all numbers of type 5, if p is odd, are horizontal numbers.

Accordingly, all numbers of type 5, if p is even, are vertical numbers.

9. Branches and Trunks of the Collatz Tree

Let's call a branch of the Collatz Tree, generated by a natural odd number a subgraph of the Collatz tree, consisting of the vertices of the Collatz tree, located strictly to the right of each other, starting from the original vertex

It is obvious that all branches consist only of horizontal numbers

Let's call a Trunk of a Collatz Tree, generated by a natural odd number a subgraph of the Collatz tree, consisting of the vertices of the Collatz tree, located strictly above each other, starting from the initial vertex

Obviously, all the trunks consist only of vertical numbers

20

10. Full trunks and full branches

Let's call the trunk of the tree full If it starts at the bottom from the vertex, below which there is a horizontal number or root number

It is obvious that all full trunks contain all vertical and only vertical numbers

Let's call a branch of the tree full If it starts to the left of the vertex, to the left of which there is a vertical number

It is obvious that all full branches contain all horizontal and only horizontal numbers

11. Direct and Reverse Vertical and Horizontal Transformations of Collatz i

Let's call the Direct Vertical Transformation of the Collatz the Transformation S, which transfers one vertical integer p to another odd natural integer$(p-1)/4$

The Direct Vertical Transformation is defined on the set of all vertical integer, that is, on the integers that can be represented as $p = 4*q+1$ (where $q = 2*k+1$, k=0,1,..)

Let's call the Reverse Vertical Transformation of the Collatz a transformation T which translates one odd positive integer p to another odd positive integer $4p+1$

The Reverse Vertical Transformation is defined on all

21

positive integers

Let's call the Direct Horizontal Transformation of a Collatz a transformation F, which translates one horizontal integer p into another odd natural integer $(3*p+1)/2$, if the integer p is a integer of type 3, 7 or 11, or into another odd natural integer $(3*p+1)/4$, if the integer p is a horizontal integer of type 1, 5 or 9

The Direct Horizontal Transformation is defined on the set of horizontal integers, that is, on the integers that cannot be represented as $p = 4*q+1$ (where $q = 2*k+1$, k=0,1,..) and which are not the root integer 1

Images of Direct Horizontal Transformation can be only integers of type 1,5,7 and 11, for example

$(3*(12*2+1)+1)/4 = 19 = 12*1+7$,
$(3*(12*1+3)+1)/2 = 23 = 12*1+11$,
$(3*(12*1+5)+1)/4 = 13 = 12*1+1$,
$(3*(12*1+7)+1)/2 = 29 = 12*2+5$,

the integers of type 3 and type 9 can never be images of a Direct Horizontal Transformation, because at any non-negative p and k the equality for t = 2 or 4 is impossible

$t*(12*p+3) != 3*(2*k+1)+1$ because $t*(12*p+3) - 3*(2*k+1) != 1$

$t*(12*p+9) != 3*(2*k+1)+1$ because $t*(12*p+9) - 3*(2*k+1) != 1$

Let's call the Reverse Horizontal Transformation of the Collatz transformation G, which translates one odd positive integer p to another odd positive integer $(2*p-1)/3$, if p is a integer of type 5 or 11, or to another positive integer (4*p-1)/3, if p is a integer of type 1 or 7

The Reverse Horizontal Transformation is defined on the set of all vertical integers and horizontal integers that are not of type 3 or 9

It is obvious that the images of the Reverse Horizontal Transformation can only be horizontal integers

It is obvious that Direct and Reverse Vertical and Horizontal Transformations of Collatz is one-to-one

12. Direct and Reverse horizontal sequence of collatz is

Let's call a Direct Horizontal Sequence of Collatz (HF), a sequence of natural integers A[k], k=0,1,...which satisfies the following condition A[k+1] = F (A[k]), which begins in any non-root natural integer or consists only of a vertical integer if it is the first in the sequence

Let's call the Reverse Horizontal Sequence of Collatz (HG) a sequence of natural integers A[k], k=0,1,...which satisfies the following condition A[k+1] = G (A[k]), which begins in any non-root natural integer or consists only of a integer of type 12*p+3 or type 12*p+9, if it is the first in the sequence

12.1 "n-Similarity" direct horizontal sequences of Collatz

If the first terms of two direct horizontal sequences differ by $12*4^n$ ($0<n$), then the first n terms of these two direct horizontal sequences have pairwise identical types

Let $12*p+q$ and $12*s+t$ be the first members of direct horizontal sequences

and $12*p+q = 12*s+t + 12*4^n$

then it is obvious that $q = t$ and $p = s + 4^n$

and consequently

$12 * p+q = 12*(s+4^n)+q$

Obviously, the natural integers $12*p+q$ and $12*s+q$ are either both vertical or both horizontal natural integers

if both natural integers $12*p+q$ and $12*s+q$ are horizontal then the images of these integers differ by (($12*3*4^n)/2$ if $q = 3,7,11$ or $(12*3*4^n) / 4$ if $q = 1,5,9$ at $n > 0$

and accordingly, the types of images of these natural integers in the direct horizontal transformation are the same types

Applying the direct horizontal transformation to the members of the above two sequences sequentially, we will get either vertical natural integers in both sequences simultaneously or horizontal natural integers at each next step

The natural integers at positions i (at $1 < i =< n$) will differ

24

by $(12*3^{(i-1)}*4^n) / v$, where $2^{(i-1)} =< v =< 4^{(i-1)}$ and therefore will have the same types

If a pair of vertical natural integers occurs before n steps, it means that both sequences have exactly the same types in pairs and have length less than n

If a pair of vertical natural integers does not occur before n steps, it means that both sequences have completely identical types in pairs at least on the first n positions

From the above it follows that any set of direct horizontal sequences, the first members of which are a series of natural integers

$12*p+q$, $12*(p+1)+q$, ..., $12*(p+4^n-1)+q$ have the same number of sequences with length greater than n for p and q, p >=0 if q = 3,5,7,9,11 or p >0 if q = 1

From this, in turn, it follows that any set of direct horizontal sequences, the first members of which are a series of natural odd natural integers

$12*p+q$, $12*p+(q+2)$, $12*p+(q+4)$, $12*p+(q+6)$, $12*p+(q+8)$, $12*p+(q+10)$, $12*(p+1)+q$, $12*(p+1)+(q+2)$, $12*(p+1)+(q+4)$, $12*(p+1)+(q+6)$, $12*(p+1)+(q+8)$, $12*(p+1)+(q+10)$,......,$12*(p+4^n-1)+q$, $12*(p+4^n-1)+(q+2)$, $12*(p+4^n-1)+(q+4)$, $12*(p+4^n-1)+(q+6)$, $12*(p+4^n-1)+(q+8)$, $12*(p+4^n-1)+(q+10)$

has the same number of sequences with length greater than n for p and q, p >=0 if q = 3,5,7,9,11 or p >0 if q = 1

12.2 "n-Similarity" reverse horizontal sequences of Collatz

If the first terms of the reverse horizontal sequences differ by $12*3^n$, then the first n terms of these sequences have pairwise identical types

Let $12*p+q$ and $12*s+t$ be the first members of the sequences and

$$12*p+q = 12*s+t + 12*3^n$$

then $q = t$ and $p = s + 3^n$ and

$$12 * p+q = 12*(s+3^n)+q$$

The natural integers $12*p+q$ and $12*s+q$ at $q = 3$ or $q = 9$ have no images in the reverse horizontal transformation (see section 11)

if both natural integers $12*p+q$ and $12*s+q$ are not 3 or 9 types then the images of these integers differ by$(12*4*3^n)/3$ if $q = 1$ and 7 or $(12*2*3^n)/3$ if $q = 5$ and11 when $n > 0$

and accordingly, the types of images of these numbers in the reverse horizontal transformation are the same types

By consistently applying the reverse horizontal transformation to the members of the above sequences, we will obtain either type 3 and 9 natural integers in both sequences simultaneously or natural integers of type 1,5,7,11 at each next step

The natural integers at positions i (at i =< n) will differ by $(12*v*3^n)/3^i$, where $2^i =< v =< 4^i$ and therefore have the same types

If a pair of natural integers of type 3 or 9 meet before n steps, it means that both sequences have completely identical types and length less than n

If a pair of natural integers of type 3 or 9 does not occur earlier than n steps, it means that both sequences have completely pairwise identical types at least on the first n positions
It follows from the above that any set of reverse horizontal sequences whose first terms are a series of natural integers

$12*p+q, 12*(p+1)+q, ..., 12*(p+3^n-1)+q$ have the same number of sequences with length greater than n for any p

From this, in turn, it follows that any set of reverse horizontal sequences, the first terms of which are a series of natural odd numbers

$12*p+q, 12*p+(q+2), 12*p+(q+4), 12*p+(q+6), 12*p+(q+8), 12*p+(q+10), 12*(p+1)+q, 12*(p+1)+(q+2), 12*(p+1)+(q+4), 12*(p+1)+(q+6), 12*(p+1)+(q+8), 12*(p+1)+(q+10),......,12*(p+3^n-1)+q, 12*(p+3^n-1)+(q+2), 12*(p+3^n-1)+(q+4), 12*(p+3^n-1)+(q+6), 12*(p+3^n-1)+(q+8), 12*(p+3^n-1)+(q+10)$

has the same number of sequences with length greater than n for any p

13. Rule (3/4)^n and rule (2/3)^n for horizontal sequences with lengths greater than n.

We assume that the sequences are arranged in some set in a row if the first members of the sequences are consecutive natural odd numbers, namely (2*k+1)+0, (2*k+1)+2, ... where k is any positive integer

The following sections will show that

the proportion of sequences whose length is greater than n in any set (whose size is a multiple of $6*4^n$) of direct horizontal sequences that are arranged in a row is $(3/4)^n$, n>0,

and that

the proportion of sequences whose lengths are greater than n in any set (the size of which is a multiple of $6*3^n$) of reverse horizontal sequences, which are arranged in a row, is $(2/3)^n$, n>0

13.1 Rule (3/4)^n for sequences whose lengths are greater than n in a set of direct horizontal sequences.

Then everywhere in section 13.1 the symbol → will denote a direct horizontal conversion F and s will denote a positive integer

13.1.1 the Rule (3/4)^1 for sequences with lengths that are greater than 1 in the set of direct horizontal sequences.

Consider direct horizontal sequences that begin in natural integers of type 3

12*0 + 3, 12*1 + 3, 12*2+ 3 and so on

all natural integers 12*p+3, p=0,1,..are horizontal and always have an image with direct horizontal conversion,

therefore, all direct horizontal sequences that begin at these numbers have a length greater than 1

Hence, in the set of direct horizontal sequences that begin in the natural integers 12*p + 3, p=0,1,2,.. all sequences are longer than 1

Consider direct horizontal sequences that begin in natural integers of type 7

12*0 + 7, 12*1 + 7, 12*2+ 7 and so on

all natural integers 12*p+7, p=0,1,..are horizontal and always have an image with direct horizontal conversion,

therefore, all direct horizontal sequences that begin at these natural integers have a length greater than 1

Hence, in the set of direct horizontal sequences that begin in the natural integers 12*p + 7, p=0,1,2,.. all sequences are longer than 1

Consider direct horizontal sequences that begin in natural

integers of type 11

12*0 + 11, 12*1 + 11, 12*2+ 11 and so on

all natural integers 12*p+11, p=0,1,..are horizontal and always have an image with direct horizontal conversion,

therefore, all direct horizontal sequences that begin at these natural integers have a length greater than 1

Hence, in the set of direct horizontal sequences that begin in the natural integers 12*p + 11, p=0,1,2,.. all sequences are longer than 1

Consider a direct horizontal sequence, which starts with natural integers like 5

12*0 + 5, 12*1 + 5, 12*2+ 5 and so on

all natural integers 12*p+5, p=0.2,.. vertical and not have the image with the direct horizontal conversion,

therefore, all direct horizontal sequences that start at these natural integers have a length of 1 and end at these natural integers.

all natural integers 12*p+5, p=1,3,.. horizontal and always have an image with direct horizontal conversion,

therefore, all direct horizontal sequences that begin at these natural integers have a length greater than 1

Hence, in the set of direct horizontal sequences that begin in the natural integers 12*p + 5, p=0,1,2,.. only half of the sequences are longer than 1

Consider a direct horizontal sequence, which starts with natural integers of the type 9

12*0 + 9, 12*1 + 9, 12*2+ 9 and so on

all natural integers 12*p+9, p=0.2,.. horizontal and always have an image with direct horizontal conversion,

therefore, all direct horizontal sequences that begin at these natural integers have a length greater than 1

all natural integers 12*p+9, p=1,3,.. vertical and not have the image with the direct horizontal conversion,

therefore, all direct horizontal sequences that start at these natural integers have a length of 1 and end at these natural integers.

Hence, in the set of direct horizontal sequences that begin in the natural integers 12*p + 9, p=0,1,2,.. only half of the sequences are longer than 1

Consider a direct horizontal sequence, which starts in the natural integers of type 1

12*1+ 1, 12*2 + 1, 12*3 + 1 and so on

all natural integers 12*p+1, p=1,3,.. vertical and not have the image with the direct horizontal conversion,

therefore, all direct horizontal sequences that start at these natural integers have a length of 1 and end at these natural integers.

31

all natural integers 12*p+1, p=2,4,.. horizontal and always have an image with direct horizontal conversion,

therefore, all direct horizontal sequences that begin at these natural integers have a length greater than 1

Hence in the set of direct horizontal sequences that begin in the natural integers 12*p + 1, p=1,2,.. only half of the sequences are longer than 1

Summing up, we see that all sequences that begin in the natural integers 12*p+3, 12*p+7, 12p+11, where p = 0,1,2,... have a length greater than 1 and half of the sequences that begin in the natural integers 12*p+5,(where p=0,1,2,...), 12*p+9 (where p=0,1,2,...), 12*p+1 (where p=1,2,3,..), have a length greater than 1

From the above reasoning and section 12.1, it follows that for any p > = 0 if q = 3,5,7,9,11, and p > 0 if q =1, and any m > 0 in the set of direct horizontal sequences

12*p+q, 12*p+(q+2), 12*p+(q+4), 12*p+(q+6), 12*p+(q+8), 12*p+(q+10), 12*(p+1)+q, 12*(p+1)+(q+2), 12*(p+1)+(q+4), 12*(p+1)+(q+6), 12*(p+1)+(q+8), 12*(p+1)+(q+10),......,12*(p+4^m-1)+q, 12*(p+4^m-1)+(q+2), 12*(p+4^m-1)+(q+4), 12*(p+4^m-1)+(q+6), 12*(p+4^m-1)+(q+8), 12*(p+4^m-1)+(q+10)

exactly $(6*4^m)*(3/4)^1$ have a length greater than 1,

namely, 18 of 24, 72 of 96 and etc

Hence (see section 12.1) the proportion of direct horizontal

32

sequences whose length is greater than 1 in any set (whose size is a multiple of 6*4^1) of direct horizontal sequences that are arranged in a row is (3/4)^1

13.1.2 Rule (3/4)^2 for sequences with lengths that are greater than 2 in the set of direct horizontal sequences.

Consider direct horizontal sequences that begin in natural integers of type 3, where p = 0,2,4,…

$12^*0 + 3 \rightarrow 12^*0 + 5,$
$12^*2 + 3 \rightarrow 12^*3 + 5,$
$12^*4 + 3 \rightarrow 12^*6 + 5,$
$12^*6 + 3 \rightarrow 12^*9 + 5,$

Note that all natural integers that stand at position number 2 in the considered horizontal lines can be represented as $12*r+5$, where $r = v*3^s+t$, at v = 0,1,2,..... and $t < 3^s$, where s = 1.

All natural integers $12*p+5$, p=0.2,.. vertical and not have the image with the direct horizontal conversion, therefore, all sequences that begin in natural integers $12*p + 3$, where p = 4 * k, k=0,1,2,... have length 2.

all natural integers $12*p+5$, p=1,3,.. horizontal and always have an image with direct horizontal transformation, so all sequences that start at $12*p + 3$, where p = 4 * k+2, k=0,1,2,... have a length greater than 2.

Consider direct horizontal sequences that begin in natural integers of type 3, where p = 1,3,5,…

33

$12^{*}1 + 3 \rightarrow 12^{*}1 + 11,$

$12^{*}3 + 3 \rightarrow 12^{*}4 + 11,$

$12^{*}5 + 3 \rightarrow 12^{*}7 + 11,$

$12^{*}7 + 3 \rightarrow 12^{*}6 + 11$

Pay attention to the fact that all the natural integers that are in the position number 2 in the direct horizontal sequence can be represented as $12*r+11$, where $r = v*3^{\wedge}s+t$, for $v = 0,1,2,....$ and $t < 3^{\wedge}s$, where $s = 1$.

All natural integers $12*p+11$, $p=1,2,3,..$ horizontal and always have an image with direct horizontal transformation, so all sequences that start at $12*p + 3$, where $p = 2 * k+1$, $k=0,1,2,...$
or $12*p + 3$, where $p = 4*k+1$, $k=0,1,2,...$ and $12 * p + 3$, where $p = 4*k+3$, $k=0,1,2, ...$ have a length greater than 2

Therefore, the proportion of direct horizontal sequences that begin at $12*p+3$ and have a length greater than 2 is ¾ of the set of sequences that begin at $12*p+3$ and whose length is greater than 1

Consider a direct horizontal sequence, which starts with natural integers of type 5 and the length of which is greater than 1

Then there are those direct horizontal sequence, which starts in a horizontal the number of type 5

$12*1 + 5$, $12*3 + 5$, $12*7+ 5$ and so on

$12^{*}1 + 5 \rightarrow 12^{*}1 + 1$

12*3 + 5 → 12*2 + 7
12*5 + 5 → 12*4 + 1
12*7 + 5 → 12*5 + 7
12*9 + 5 → 12*7 + 1
12*11 + 5 → 12*8 + 7
12*13 + 5 → 12*10 + 1
12*15 + 5 → 12*11 + 7

Consider the sequence that starts at natural integers 12*p + 5 where p = 4*k+1, k=0,1,2,...

12*1 + 5, 12*5 + 5, 12*9+ 5, 12*13 + 5 and so on

12*1 + 5 → 12*1 + 1
12*5 + 5 → 12*4 + 1
12*9 + 5 → 12*7 + 1
12*13 + 5 → 12*10 + 1

Note that all natural integers that stand at position number 2 in the considered horizontal lines can be represented as 12*r+1, where r = v*3^s+t, at v = 0,1,2,..... and t < 3^s, where s = 1.

All natural integers 12*p+1, p=2*k+1, k=0,1,2,.. vertical and not have the image with the direct horizontal conversion,
therefore, all sequences that begin in natural integers 12*p + 5, where p = 8 * k+1, k=0,1,2,... have length 2.

all natural integers 12*p+1, p=2*k, k=0,1,2,.. are horizontal and have an image with direct horizontal transformation, so all sequences that start at in natural integers 12*p+5, where p = 8*k+5 have a length greater than 2

Consider sequences that start at 12*p + 5, where p = 4 * k+3, k=0,1,2,...

12*3 + 5, 12*7 + 5, 12*11+ 5 and so on

12*3 + 5 → 12*2 + 7
12*7 + 5 → 12*5 + 7
12*11 + 5 → 12*8 + 7
12*15 + 5 → 12*11 + 7

Pay attention to the fact that all the natural integers that are in the position number 2 in the direct horizontal sequence can be represented as 12*r+7, where r = v*3^s+t, for v = 0,1,2,.... and t < 3^s, where s = 1.

All natural integers 12*p+7, p= 0,1,2,.. horizontal and always have an image with direct horizontal transformation, so all direct horizontal sequences that start at 12*p + 5, where p = 4 * k+3, k=0,1,2,... have a length greater than 2.
Or 12*p + 5, where p = 8 * k+3, k=0,1,2,... and 12*p + 5, where p = 8 * k+7, k=0,1,2,...

Therefore, the proportion of direct horizontal sequences that begin at 12*p+5 and have a length greater than 2 is ¾ of the set of direct horizontal sequences that begin at 12*p+5 and whose length is greater than 1,

namely this number, 12*p+5 (p=5,13,.or 8*k+5, k=0,1,2,...), 12*p+5 (p=3,11,19,.or 8*k+3, k=0,1,2,...), 12*p+5 (p=7,15,23,.or 8*k+7, k=0,1,2,...)

Consider a direct horizontal sequence, which starts with
natural integers of type 7, where p = 1,3,5,...

12*1 + 7 → 12*2 + 5,
12*3 + 7 → 12*5 + 5,
12*5 + 7 → 12*8 + 5,
12*7 + 7 → 12*11 + 5,

Note that all natural integers that stand at position number 2
in the considered horizontal lines can be represented as
12*r+5, where r = v*3^s+t, at v = 0,1,2,..... and t < 3^s,
where s = 1.

All natural integers 12*p+5, p=0.2,.. vertical and not have
the image with the direct horizontal conversion,
therefore, all direct horizontal sequence, which starts in
natural integers 12*p + 7, where p = 4*k+1, k=0,1,2,...
have length 2.

all natural integers 12*p+5, p=1,3,.. horizontal and always
have an image with direct horizontal transformation, so all
sequences that start at 12*p + 7, where p = 4 * k+3,
k=0,1,2,... have a length greater than 2.

Consider a direct horizontal sequence, which starts with
natural integers of type 7, where p = 0,2,4,...

12*0 + 7 → 12*0 + 11,
12*2 + 7 → 12*3 + 11,
12*4 + 7 → 12*6 + 11,
12*6 + 7 → 12*9 + 11,

Pay attention to the fact that all the natural integers that are
in the position number 2 in the direct horizontal sequence
can be represented as 12*r+11, where r = v*3^s+t, for v =

0,1,2,.... and t < 3^s, where s = 1.

all natural integers 12*p+11, p=1,2,3,.. horizontal and always have an image with direct horizontal transformation, so all direct horizontal sequences that start at 12*p + 7, where p = 2 * k, k=0,1,2,... have a length greater than 2.
Or 12 * p + 7, where p = 4*k, k=0,1,2,... and 12 * p + 7, where p = 4*k+2, k=0,1,2,...

Therefore, the proportion of direct horizontal sequences that begin at 12*p+7 and have a length greater than 2 is ¾ of the set of direct horizontal sequences that begin at 12*p+7 and whose length is greater than 1,

namely this number 12 * p+7 (p=0,4,8,...or p=4*k), 12*p+7 (p=2,6,10,..or p=4*k+2), 12 * p+7 (p=3,7,11, ... or p=4*k+3)

Consider a direct horizontal sequence, which starts with natural integers of the type 9 and the length of which is greater than 1

Then there are those direct horizontal sequence, which starts in a horizontal the number of type 9

12*0 + 9, 12*2+ 9, 12*4+9 and so on

12*0 + 9 → 12*0 + 7
12*2 + 9 → 12*2 + 1
12*4 + 9 → 12*3 + 7
12*6 + 9 → 12*5 + 1
12*8 + 9 → 12*6 + 7

12*10 + 9 → 12*8 + 1
12*12 + 9 → 12*9 + 7
12*14 + 9 → 12*11 + 1

Consider a direct horizontal sequence, which starts in natural integers 12*p + 9 where p = 4*k, k=0,1,2,...

12*0 + 9, 12*4 + 9, 12*8+ 9 and so on

12*0 + 9 → 12*0 + 7
12*4 + 9 → 12*3 + 7
12*8 + 9 → 12*6 + 7
12*12 +9 → 12*9 + 7

Pay attention to the fact that all the natural integers that are in the position number 2 in the direct horizontal sequence can be represented as 12*r+7, where r = v*3^s+t, for v = 0,1,2,.... and t < 3^s, where s = 1.

All natural integers 12*p+7, p= 0,1,2,.. horizontal and always have an image with direct horizontal transformation, so all sequences that start at the natural integers 12*p + 9, where p = 4 * k, k=0,1,2,... have a length greater than 2.
Or 12 * p + 9, where p = 8*k, k=0,1,2,... and 12 * p + 9, where p = 8*k+4, k=0,1,2,...

Consider a direct horizontal sequence, which starts in natural integers 12*p + 9 where p = 4*k+2, k=0,1,2,...

12*2 + 9, 12*6 + 9, 12*10+ 9, 12*14 + 9 and so on

12*2 + 9 → 12*2 + 1

$12*6 + 9 \rightarrow 12*5 + 1$

$12*10 + 9 \rightarrow 12*8 + 1$

$12*14 + 9 \rightarrow 12*11 + 1$

Note that all natural integers that stand at position number 2 in the considered horizontal lines can be represented as $12*r+1$, where $r = v*3\text{^}s+t$, at $v = 0,1,2,.....$ and $t < 3\text{^}s$, where $s = 1$.

All natural integers $12*p+1$, $p=2*k+1$, $k=0,1,2,..$ vertical and not have the image with the direct horizontal conversion,
therefore, all direct horizontal sequences that begin at the natural integers $12*p + 9$, where $p = 8 * k+6$, $k=0,1,2,...$ have length 2.

all natural integers $12*p+1$, $p=2*k$, $k=1,2,..$ are horizontal and have an image with direct horizontal transformation, so all direct horizontal sequences that start at in natural integers $12*p+9$, where $p = 8*k+2$ have a length greater than 2

Therefore, the proportion of direct horizontal sequences that begin at $12*p+9$ and have a length greater than 2 is ¾ of the set of direct horizontal sequences that begin at $12*p+9$ and whose length is greater than 1,

namely, these are the natural integers $12*p + 9$ (p=0.8.16,.. or p = 8*k, k=0,1,2,...12 * p+9), $12*p+9$ (p=2,10,18,...or 8*k+2, k=0,1,2,...), $12*p+9$ (p=4,12,20,... or 8*k+4, k=0,1,2,..)

Consider direct horizontal sequences that begin in natural

integers of type 11, where p = 1,3,5,...

12*1 + 11 → 12*2 + 11,
12*3 + 11 → 12*5 + 11,
12*5 + 11 → 12*8 + 11,
12*7 + 11 → 12*11 + 11,

Pay attention to the fact that all the natural integers that are in the position number 2 in the direct horizontal sequence can be represented as $12*r+11$, where $r = v*3^s+t$, for v = 0,1,2,.... and $t < 3^s$, where s = 1.

All natural integers $12*p+11$, p=1,2,3,.. horizontal and always have an image with direct horizontal transformation, so all direct horizontal sequences that start at $12*p + 11$, where p = 2 * k+1, k=0,1,2,... have a length greater than 2.

Consider direct horizontal sequences that begin in natural integers of type 11, where p = 0,2,4,...

12*0 + 11 → 12*1 + 5,
12*2 + 11 → 12*4 + 5,
12*4 + 11 → 12*7 + 5,
12*6 + 11 → 12*10 + 5,

Note that all natural integers that stand at position number 2 in the considered horizontal lines can be represented as $12*r+5$, where $r = v*3^s+t$, at v = 0,1,2,...... and $t < 3^s$, where s = 1.

All natural integers $12*p+5$, p=0.2,.. vertical and not have the image with the direct horizontal conversion, therefore, all direct horizontal sequences that begin at the natural integers 12*p + 11, where p = 4 * k+2, k=0,1,2,... have length 2.

all natural integers 12*p+5, p=1,3,.. horizontal and always have an image with direct horizontal conversion, so all direct horizontal sequence, which starts in natural integers 12*p + 11, where p = 4*k, k=0,1,2,... have a length greater than 2.

Therefore, the proportion of direct horizontal sequences that begin at 12*p+11 and have a length greater than 2 is ¾ of the set of direct horizontal sequences that begin at 12*p+11 and whose length is greater than 1,

namely this number 12*p+11 (p=0,4,18,..or p=4*k), 12 * p+11 (p=1,5,9,... or p=4*k+1), 12 * p+11 (p=3,7,11, ... or p=4*k+3)

Consider a direct horizontal sequence, which starts in the natural integers of type 1 and the length of which is greater than 1

That is, those sequences that begin in horizontal natural integers of type 1

12*2 + 1, 12*4 + 1, 12*6 + 1 and so on

12*2 + 1 → 12*1 + 7
12*4 + 1 → 12*3 + 1
12*6 + 1 → 12*4 + 7
12*8 + 1 → 12*6 + 1
12*10 + 1 → 12*7 + 7
12*12 + 1 → 12*9 + 1

12*14 + 1 → 12*10 + 7
12*16 + 1 → 12*12 + 1

Consider sequences that start at 12*p + 1, where p = 4 * k+2, k=0,1,2,...

12*2 + 1, 12*6 + 1, 12*10+ 1 and so on

12*2 + 1 → 12*1 + 7
12*6 + 1 → 12*4 + 7
12*10 + 1 → 12*7 + 7
12*14 + 1 → 12*10 + 7

Pay attention to the fact that all the natural integers that are in the position number 2 in the direct horizontal sequence can be represented as $12*r+7$, where $r = v*3^s+t$, for v = 0,1,2,.... and $t < 3^s$, where s = 1.

All natural integers $12*p+7$, p= 0,1,2,.. horizontal and always have an image under direct horizontal transformation, so all direct horizontal sequences that start at 12*p + 1, where p = 4 * k+2, k=0,1,2,... have a length greater than 2.

Consider a direct horizontal sequence, which starts in natural integers 12*p + 1 where p = 4*k, k=1,2,3,...

12*4 + 1, 12*8 + 1, 12*12+ 1, 12*16 + 1 and so on

12*4 + 1 → 12*3 + 1
12*8 + 1 → 12*6 + 1
12*12 + 1 → 12*9 + 1

43

$12*16 + 1 \rightarrow 12*12 + 1$

Note that all natural integers that stand at position number 2 in the considered horizontal lines can be represented as $12*r+1$, where $r = v*3^s+t$, at $v = 0,1,2,.....$ and $t < 3^s$, where $s = 1$.

All natural integers $12*p+1$, $p=2*k+1$, $k=0,1,2,..$ vertical and not have the image with the direct horizontal conversion,
therefore, all direct horizontal sequence, which starts in natural integers $12*p + 1$, where $p = 8*k+4$, $k=0,1,2,...$ have length 2.

all natural integers $12*p+1$, $p=2*k$, $k=1,2,..$ are horizontal and have an image under direct horizontal transformation, so all direct horizontal sequences that start at in natural integers $12*p+1$, where $p = 8*k$ have a length greater than 2

Therefore, the proportion of direct horizontal sequences that begin at $12*p+1$ and have a length greater than 2 is ¾ of the set of direct horizontal sequences that begin at $12*p+1$ and whose length is greater than 1,

namely this number $12*p + 1$ (p=0,8,16.. or $p = 8*k$, k=0,1,2,...12 * p+9), $12*p+1$ (p=2,10,18,...or $8*k+2$, k=0,1,2,...), $12*p+1$ (p=6,14,22,... or $8*k+6$, k=0,1,2,..)

From the above reasoning and section 12.1, it follows that for any $p >= 0$ if $q = 3,5,7,9,11$, and $p > 0$ if $q =1$, and any $m > 0$ in the set of direct horizontal sequences

$12*p+q$, $12*p+(q+2)$, $12*p+(q+4)$, $12*p+(q+6)$, $12*p+(q+8)$, $12*p+(q+10)$,
$12*(p+1)+q$, $12*(p+1)+(q+2)$, $12*(p+1)+(q+4)$,

12*(p+1)+(q+6), 12*(p+1)+(q+8), 12*(p+1)+(q+10),......,12*(p+4*4^m-1)+q, 12*(p+4*4^m-1)+(q+2), 12*(p+4*4^m-1)+(q+4), 12*(p+4 * 4^m-1)+(q+6), 12*(p+4*4^m-1)+(q+8), 12*(p+4*4^m-1)+(q+10)

exactly (6*4*4^m)*(3/4)^2 are longer than 2,

namely 54 of 96, 216 of 384 and etc

Hence (see section 12.1) the proportion of direct horizontal sequences whose length is greater than 2 in any set (the size of which is a multiple 6*4*4^2) direct horizontal sequences, which are arranged in a row, equal to (3/4)^2

13.1.3 Rule (3/4)^n for sequences whose lengths are greater than n in a set of direct horizontal sequences.

Members of direct horizontal sequences starting from the second position can be only natural integers of the 5th, 7th, 11th and 1st types.

The natural integers of 3-th and 9-th types can't be images with direct horizontal transformation (see section 11).

In the following sections from 13.1.3.1 to 13.1.3.5 it will be proved that the proportion of direct horizontal sequences whose length is greater than n is (¾)^n of the set of all direct horizontal sequences

13.1.3.1 Direct horizontal sequences in which the members n, n>1 are natural integers of type 5

Consider direct horizontal sequences that begin in terms of 12*p+q and in which the terms at position n, n>1 are natural integers of type 5

$$12^*p+q \rightarrow .. \rightarrow 12^*r+5$$

let r = v*3^s+t, where v = 0,1,2,.... and t < 3^s, s>0,

then

$$12*(v*3^s+t)+5 = 12 * v*3^s+12 * t+5 = 4*(v*3^{(s+1)}+3*t+1)+1$$

therefore, if the natural integer v is even, then for any even natural integer t, the natural integer 4*(v*3^(s+1)+3*t+1)+1 will be vertical 4*(2*k+1)+1 and therefore this natural integer will be the last in a direct horizontal sequence that begins in the natural integer 12*(v*3^s+t)+5 and, respectively, in the natural integer 12*p+q,

at the same time, if the natural integer v is even, then for any odd natural integer t, the natural integer 4*(v*3^(s+1)+3 * t+1)+1 will be horizontal 4*(2*k)+1 and therefore this natural integer will not be the last in a direct horizontal sequence that begins in the natural integer 12*(v*3^s+t)+5 and, respectively, in the natural integer 12*p+q,

further, if the natural integer V is odd, then for any even natural integer t the natural integer 4*(v*3^(s+1)+3*t+1)+1 will be horizontal 4*(2*k)+1 and therefore will not be the last in a direct horizontal sequence that begins in the natural integer 12*(v*3^s+t)+5 and, respectively, in the natural integer 12 * p+q,

at the same time, if the natural integer v is odd, then for any odd natural integer t the natural integer 4*(v*3^(s+1)+3*t+1)+1 will be vertical 4*(2*k+1)+1 and therefore the last in a direct horizontal sequence that begins in the natural integer 12*(v*3^s+t)+5 and, respectively, in the natural integer 12*p+q.

Therefore, if v = 0,1,2,.... then whatever non-negative integers t are, exactly half of the sequences that begin at 12*p+q and have the element 12*(v*3^s+t) +5 in place of n end at this element and exactly half of the sequences have the next element at n+1

Let's consider what these following elements can be

The following pairs of natural integers v and t are possible, for which the direct horizontal sequences are continued

v=4*y+0, t =4*z+1,
v=4*y+0, t =4*z+3,
v=4*y+2, t =4*z+1,
v=4*y+2, t =4*z+3,
v=4*y+1, t =4*z+0,
v=4*y+1, t =4*z+2,
v=4*y+3, t =4*z+0,
v=4*y+3, t =4*z+2

Let v=4*y+0, y = 0,1,2,...., and t =4*z+1 then

$12^*(v^*3^\wedge s+t)+5 = 12^*v^*3^\wedge s+12^*t+5 \rightarrow$
$(3^*(12^*v^*3^\wedge s+12^*t+5)+1)/4 =$

$(12*((4*y+0)*3^{(s+1)}+(4*z+1)*3)+5*3+1)/4 =$
$(12*((4*y)*3^{(s+1)}+(4*z)*3)+12*3+5*3+1)/4 =$
$12*(y*3^{(s+1)}+z*3)+13 =$
$12*(y*3^{(s+1)}+(z*3+1))+1$

Let us note that

$z*4+1 < 3^s,$
$z*3+1 < 3^s$
$z*3+1 < 3^{(s+1)}$

Therefore if $v=4*y+0$, $y = 0,1,2,....$, and $t = 4*z+1$, then

$12*(v*3^s+t)+5 \rightarrow 12*(y*3^{(s+1)}+(z*3+1))+1$

this means that the images of natural integers of type 5 $(12*(v*3^s+t)+5)$ in this case will be natural integers of type 1 $(12*(y*3^{(s+1)}+(z*3+1))+1)$, which can be represented as $12*r+1$ and $r = v*3^{(s+1)}+t$, when $v = 0,1,2,....$ and $t < 3^{(s+1)}$, $s = 1,2,3,4,...$

Let $v=4*y+0$, $y = 0,1,2,....$, and $t = 4*z+3$ then

$12*(v*3^s+t)+5 = 12*v*3^s+12*t+5 \rightarrow$
$(3*(12*v*3^s+12*t+5)+1)/4 =$
$(12*((4*y+0)*3^{(s+1)}+(4*z+3)*3)+5*3+1)/4 =$
$(12*((4*y+0)*3^{(s+1)}+(4*z)*3)+12*3*3+5*3+1)/4 =$
$12*(y*3^{(s+1)}+z*3)+31 =$
$12*(y*3^{(s+1)}+(z*3+2))+7$

Let us note that

$z*4+3 < 3^s,$
$z*3+2 < 3^s,$

48

$z*3+2 < 3^\wedge(s+1)$

Therefore if v=4*y+0, y = 0,1,2,...., and t =4*z+3, then

$12^*(v^*3^\wedge s+t)+5 \rightarrow 12^*(y^*3^\wedge(s+1)+(z^*3+2))+7$

this means that the images of natural integers of type 5 (12*(v*3^s+t)+5) in this case will be natural integers of type 7 (12*(y*3^(s+1)+(z*3+2))+7), which can be represented as 12*r+7 and r = v*3^(s+1)+t, when v = 0,1,2,.... and t < 3^(s+1), s = 1,2,3,4,...

Let v=4*y+2, y = 0,1,2,...., and t =4*z+1 then

$12^*(v^*3^\wedge s+t)+5 = 12^*v^*3^\wedge s+12^*t+5 \rightarrow$
$(3^*(12^*v^*3^\wedge s+12^*t+5)+1)/4 =$
$(12^*((4^*y+2)^*3^\wedge(s+1)+(4^*z+1)^*3)+5^*3+1)/4 =$
$(12^*((4^*y)^*3^\wedge(s+1)+(4^*z)^*3)+12^*2^*3^\wedge(s+1)+$
$12^*3+5^*3+1)/4 =$
$12^*(y^*3^\wedge(s+1)+z^*3)+6^*3^\wedge(s+1)+13$

natural integers 3^s can be represented as

$3^\wedge s = 12^*m1 + 3, s = 1,3,5,...$
or
$3^\wedge s = 12^*m2 + 9, s = 2,4,6,...$

accordingly natural integers 6*3^(s+1) can be represented as

$6^*3^\wedge(s+1) = 12^*m3 + 6, s = 1,3,5,...$
or
$6^*3^\wedge(s+1) = 12^*m4 + 6, s = 2,4,6,...$

accordingly natural integers 6*3^(s+1)+13 can be represented as

6*3^(s+1)+13 = 12*m5 + 7, s = 1,3,5,...
or
6*3^(s+1)+13 = 12*m6 + 7, s = 2,4,6,...

accordingly

12*(y*3^(s+1)+z*3)+6*3^(s+1)+13 =
12*(y*3^(s+1)+z*3+m5)+7, s = 1,3,5,...
or
12*(y*3^(s+1)+z*3)+6*3^(s+1)+13 =
12*(y*3^(s+1)+z*3+m6)+7, s = 2,4,6,...

Let us note that

m5 = (6*3^(s+1)+6)/12 = (3^(s+1)+1)/2, s = 1,3,5,...

4*z+1 < 3^s,
3*z < 3^(s+1)/3 - 1,
3*z < 3^(s+1)/2 - 1,
(3^(s+1)+1)/2 + 3*z < 3^(s+1)/2+ 1/2 + 3^(s+1)/2 - 1
(3^(s+1)+1)/2 + z*3 < 3^(s+1) – 1/2
(3^(s+1)+1)/2 + z*3 < 3^(s+1)

and

m6 = (6*3^(s+1)+6)/12 = (3^(s+1)+1)/2, s = 2,4,6,...

4*z+1 < 3^s,
3*z < 3^(s+1)/3 - 1,
3*z < 3^(s+1)/2 - 1,
(3^(s+1)+1)/2 + 3*z < 3^(s+1)/2+ 1/2 + 3^(s+1)/2 - 1
(3^(s+1)+1)/2 + z*3 < 3^(s+1) – 1/2
(3^(s+1)+1)/2 + z*3 < 3^(s+1)

Therefore if v=4*y+2, y = 0,1,2,…., and t =4*z+1, then

$12^*(v^*3\text{^}s+t)+5 \rightarrow 12^*(y^*3\text{^}(s+1)+z^*3+(3\text{^}(s+1)+1)/2)+7$

this means that the images of natural integers of type 5 (12*(v*3^s+t)+5) in this case will be natural integers of type 7 (12*(y*3^(s+1)+z*3+(3^(s+1)+1)/2)+7), which can be represented as 12*r+7 and r = v*3^(s+1)+t, when v = 0,1,2,…. and t < 3^(s+1), s = 1,2,3,4,…

Let v=4*y+2, y = 0,1,2,…., and t =4*z+3 then

$12^*(v^*3\text{^}s+t)+5 = 12^*v^*3\text{^}s+12^*t+5 \rightarrow$
$(3^*(12^*v^*3\text{^}s+12^*t+5)+1)/4 =$
$(12^*((4^*y+2)^*3\text{^}(s+1)+(4^*z+3)^*3)+5^*3+1)/4 =$
$(12^*((4^*y)^*3\text{^}(s+1)+(4^*z)^*3))+12^*2^*3\text{^}(s+1)+$
$12^*3^*3+5^*3+1)/4 =$
$12^*(y^*3\text{^}(s+1)+z^*3)+6^*3\text{^}(s+1)+31$

natural integers 3^s can be represented as

$3\text{^}s = 12^*m1 + 3, s = 1,3,5,…$
or
$3\text{^}s = 12^*m2 + 9, s = 2,4,6,…$

 accordingly natural integers 6*3^(s+1) can be represented as

$6^*3\text{^}(s+1) = 12^*m3 + 6, s = 1,3,5,…$
or
$6^*3\text{^}(s+1) = 12^*m4 + 6, s = 2,4,6,…$

 accordingly natural integers 6*3^(s+1)+31 can be

represented as

$6*3^{(s+1)}+31 = 12*m5 + 1$, $s = 1,3,5,\ldots$
or
$6*3^{(s+1)}+31 = 12*m6 + 1$, $s = 2,4,6,\ldots$

accordingly

$12*(y*3^{(s+1)}+z*3)+6*3^{(s+1)}+31 =$
$12*(y*3^{(s+1)}+z*3+m5)+1$, $s = 1,3,5,\ldots$
or
$12*(y*3^{(s+1)}+z*3)+6*3^{(s+1)}+31 =$
$12*(y*3^{(s+1)}+z*3+m6)+1$, $s = 2,4,6,\ldots$

Let us note that

$m5 = (6*3^{(s+1)}+30)/12 = (3^{(s+1)}+5)/2$, $s = 1,3,5,\ldots$

$4*z+1 < 3^s$,
$4*z < 3^s - 1$,
$4*z =< 3^s - 1-2$,
$3*z =< 3^{(s+1)}/4 - 9/4$,
$1 < 3^{(s+1)}$,
$0 < (3^{(s+1)} - 1)/4$,
$(3^{(s+1)}+5)/2 + 3*z < 3^{(s+1)}/2+ 10/4 + 3^{(s+1)}/4 - 9/4 + 3^{(s+1)}/4 - 1/4$,
$(3^{(s+1)}+5)/2+z*3 < 3^{(s+1)}$

and

$m6 = (6*3^{(s+1)}+30)/12 = (3^{(s+1)}+5)/2$, $s = 2,4,6,\ldots$

$4*z+1 < 3^s$,
$4*z < 3^s - 1$,
$4*z =< 3^s - 1-2$,
$3*z =< 3^{(s+1)}/4 - 9/4$,

$1 < 3^{\wedge}(s+1),$

$0 < (3^{\wedge}(s+1) - 1)/4,$

$(3^{\wedge}(s+1)+5)/2 + 3*z < 3^{\wedge}(s+1)/2 + 10/4 + 3^{\wedge}(s+1)/4 - 9/4 + 3^{\wedge}(s+1)/4 - 1/4,$

$(3^{\wedge}(s+1)+5)/2 + z*3 < 3^{\wedge}(s+1)$

Therefore if v=4*y+2, y = 0,1,2,...., and t =4*z+3, then

$12^{*}(v*3^{\wedge}s+t)+5 \rightarrow 12^{*}(y*3^{\wedge}(s+1)+z*3+(3^{\wedge}(s+1)+5)/2)+1$

this means that the images of natural integers of type 5 (12*(v*3^s+t)+5) in this case will be natural integers of type 1 (12*(y*3^(s+1)+z*3+(3^(s+1)+5)/2)+1), which can be represented as 12*r+1 and r = v*3^(s+1)+t, when v = 0,1,2,.... and t < 3^(s+1), s = 1,2,3,4,...

Let v=4*y+1, y = 0,1,2,...., and t =4*z+0 then

$12^{*}(v*3^{\wedge}s+t)+5 = 12^{*}v*3^{\wedge}s+12^{*}t+5 \rightarrow$
$(3^{\wedge}(12^{*}v*3^{\wedge}s+12^{*}t+5)+1)/4 =$
$(12*((4*y+1)*3^{\wedge}(s+1)+(4*z+0)*3)+5*3+1)/4 =$
$(12*((4*y)*3^{\wedge}(s+1)+(4*z)*3)+12*3^{\wedge}(s+1)+5*3+1)/4 =$
$12*(y*3^{\wedge}(s+1)+z*3)+3*3^{\wedge}(s+1)+4$

natural integers 3^s can be represented as

$3^{\wedge}s = 12*m1 + 3, s = 1,3,5,...$
or
$3^{\wedge}s = 12*m2 + 9, s = 2,4,6,...$

 accordingly natural integers 3*3^(s+1) can be represented as

$3*3^{\wedge}(s+1) = 12*m3 + 3, s = 1,3,5,...$

or
$3*3^{(s+1)} = 12*m4 + 9$, s = 2,4,6,…

 accordingly natural integers $3*3^{(s+1)}+4$ can be represented as

$3*3^{(s+1)}+4 = 12*m5 + 7$, s = 1,3,5,…
or
$3*3^{(s+1)}+4 = 12*m6 + 1$, s = 2,4,6,…

 accordingly

$12*(y*3^{(s+1)}+z*3)+3*3^{(s+1)}+4 =$
$12*(y*3^{(s+1)}+z*3+m5)+7$, s = 1,3,5,…
or
$12*(y*3^{(s+1)}+z*3)+3*3^{(s+1)}+4 =$
$12*(y*3^{(s+1)}+z*3+m6)+1$, s = 2,4,6,…

Let us note that

$m5 = (3*3^{(s+1)}-3)/12 = (3^{(s+1)}-1)/4$, s = 1,3,5,…

$4*z < 3^s$,
$3*z < 3^{(s+1)}/4$,
$(3^{(s+1)}-1)/4+z*3 < 3^{(s+1)}/4 -1/4 + 3^{(s+1)}/4$
$(3^{(s+1)}-1)/4+z*3 < 3^{(s+1)}/2$
$(3^{(s+1)}-1)/4+z*3 < 3^{(s+1)}$

and

$m6 = (3*3^{(s+1)}+3)/12 = (3^{(s+1)}+1)/4$, s = 2,4,6,…

$4*z < 3^s$,
$4*z =< 3^s - 1$,
$3*z =< 3^{(s+1)}/4 – 3/4$,
$(3^{(s+1)}+1)/4 + 3*z =< 3^{(s+1)}/4+ 1/4 + 3^{(s+1)}/4 – 3/4$,

54

$(3^{\wedge}(s+1)+1)/4+z*3 < 3^{\wedge}(s+1)/2 - 1/2$

$(3^{\wedge}(s+1)+1)/4+z*3 < 3^{\wedge}(s+1)$

Therefore if $v=4*y+1$, $y = 0,1,2,....$, and $t =4*z+0$, then

$12^{*}(v^{*}3^{\wedge}s+t)+5 \rightarrow 12^{*}(y^{*}3^{\wedge}(s+1)+z^{*}3+(3^{\wedge}(s+1)-1)/4)+7$, $s = 1,3,5,...$

or

$12^{*}(v^{*}3^{\wedge}s+t)+5 \rightarrow 12^{*}(y^{*}3^{\wedge}(s+1)+z^{*}3+(3^{\wedge}(s+1)+1)/4)+1$, $s = 2,4,6,...$

this means that the images of natural integers of type 5 $(12*(v*3^{\wedge}s+t)+5)$ in this case will be natural integers of type 7 $(12*(y*3^{\wedge}(s+1)+z*3+(3^{\wedge}(s+1)-1)/4)+7)$, which can be represented as $12*r+7$ and $r = v*3^{\wedge}(s+1)+t$, when $v = 0,1,2,....$ and $t < 3^{\wedge}(s+1)$, $s = 1,3,5,...$

or means that

the images of natural integers of type 5 $(12*(v*3^{\wedge}s+t)+5)$ in this case will be natural integers of type 1 $(12*(y*3^{\wedge}(s+1)+z*3+(3^{\wedge}(s+1)+1)/4)+1)$, which can be represented as $12*r+1$ and $r = v*3^{\wedge}(s+1)+t$, when $v = 0,1,2,....$ and $t < 3^{\wedge}(s+1)$, $s = 2,4,6,...$

Let $v=4*y+1$, $y = 0,1,2,....$, and $t =4*z+2$ then

$12^{*}(v^{*}3^{\wedge}s+t)+5 = 12^{*}v^{*}3^{\wedge}s+12^{*}t+5 \rightarrow$

$(3^{*}(12^{*}v^{*}3^{\wedge}s+12^{*}t+5)+1)/4 =$

$(12*((4*y+1)*3^{\wedge}(s+1)+(4*z+2)*3)+5*3+1)/4 =$

$(12*((4*y)*3^{\wedge}(s+1)+(4*z)*3))+12*3^{\wedge}(s+1)+12*2*3+5*3+1)/4 =$

$12*(y*3^{\wedge}(s+1)+z*3)+3*3^{\wedge}(s+1)+18+4$

natural integers 3^s can be represented as

in the form $3^s = 12*m1 +3$, when s = 1,3,5,...,
and
in the form $3^s = 12*m2 +9$, when s = 2,4,6,...

accordingly natural integers $3*3^{(s+1)}$ can be represented as

in the form $3*3^{(s+1)} = 12*m3 +3$, when s = 1,3,5,...,
and
in the form $3*3^{(s+1)} = 12*m4 +9$, when s = 2,4,6,...

accordingly natural integers $3*3^{(s+1)}+18+4$ can be represented as

in the form $3*3^{(s+1)}+18+4 = 12*m5 +1$, when s = 1,3,5,...,
and
in the form $3*3^{(s+1)}+18+4 = 12*m6 +7$, when s = 2,4,6,...

accordingly

$12*(y*3^{(s+1)}+z*3)+3*3^{(s+1)}+18+4 =$
$12*(y*3^{(s+1)}+z*3+m5)+1$, s = 1,3,5,...,
or
$12*(y*3^{(s+1)}+z*3)+3*3^{(s+1)}+18+4 =$
$12*(y*3^{(s+1)}+z*3+m6)+7$, s = 2,4,6,...

Let us note that

$m5 = (3*3^{(s+1)}+21)/12 = (3^{(s+1)}+7)/4$, s = 1,3,5,...,

$4*z+2 < 3^s$,
$4*z < 3^s -2$,

$4*z =< 3^s - 2-1$,

$3*z =< 3^{(s+1)}/4 - 9/4$,

$(3^{(s+1)}+7)/4 + 3*z =< 3^{(s+1)}/4 + 7/4 + 3^{(s+1)}/4 - 9/4$,

$(3^{(s+1)}+7)/4 + 3*z =< 3^{(s+1)}/2-1/2$,

$(3^{(s+1)}+7)/4+z*3 < 3^{(s+1)}$

and

$m6 = (3*3^{(s+1)}+15)/12 = (3^{(s+1)}+5)/4$, $s = 2,4,6,…$

$4*z+2 < 3^s$,

$4*z < 3^s -2$,

$3*z < 3^{(s+1)}/4 - 6/4$,

$(3^{(s+1)}+5)/4 + 3*z < 3^{(s+1)}/4 + 5/4 + 3^{(s+1)}/4 - 6/4$,

$(3^{(s+1)}+5)/4+z*3 < 3^{(s+1)}$

Therefore if $v=4*y+1$, $y = 0,1,2,….$, and $t =4*z+2$, then

$12*(v*3^s+t)+5 \rightarrow 12*(y*3^{(s+1)}+z*3+(3^{(s+1)}+7)/4) +1$, $s = 1,3,5,…$

or

$12*(v*3^s+t)+5 \rightarrow 12*(y*3^{(s+1)}+z*3+(3^{(s+1)}+5)/4) +7$, $s = 2,4,6,…$

this means that the images of natural integers of type 5 $(12*(v*3^s+t)+5)$ in this case will be natural integers of type 1 $(12*(y*3^{(s+1)}+z*3+(3^{(s+1)}+7)/4) +1)$, which can be represented as $12*r+1$ and $r = v*3^{(s+1)}+t$, when $v = 0,1,2,….$ and $t < 3^{(s+1)}$, $s = 1,3,5,…$

or this means that the images of natural integers of type 5 $(12*(v*3^s+t)+5)$ in this case will be natural integers of type 7 $(12*(y*3^{(s+1)}+z*3+(3^{(s+1)}+5)/4) +7)$, which can be represented as $12*r+7$ and $r = v*3^{(s+1)}+t$, when $v = 0,1,2,….$ and $t < 3^{(s+1)}$, $s = 2,4,6,…$

Let v=4*y+3, y = 0,1,2,...., and t =4*z+0 then

12*(v*3^s+t)+5 = 12*v*3^s+12*t+5 →
(3*(12*v*3^s+12*t+5)+1)/4 =
(12*((4*y+3)*3^(s+1)+(4*z+0)*3)+5*3+1)/4 =
(12*((4*y)*3^(s+1)+(4*z)*3)+12*3*3^(s+1)+5*3+1)/4 =
12*(y*3^(s+1)+z*3)+3*3*3^(s+1)+4 =

natural integers 3^s can be represented as

in the form 3^s = 12*m1 +3, when s = 1,3,5,...,
and
in the form 3^s = 12*m2 +9, when s = 2,4,6,...

 accordingly natural integers 3*3*3^(s+1) can be
represented as

in the form 3*3*3^(s+1) = 12*m3 +9, when s = 1,3,5,...,
and
in the form 3*3*3^(s+1) = 12*m4 +3, when s = 2,4,6,...

 accordingly natural integers 3*3*3^(s+1)+4 can be
represented as

in the form 3*3*3^(s+1)+4 = 12*m5 +1, when s = 1,3,5,...,
and
in the form 3*3*3^(s+1)+4 = 12*m6 +7, when s = 2,4,6,...

 accordingly

12*(y*3^(s+1)+z*3)+3*3*3^(s+1)+4 =
12*(y*3^(s+1)+z*3+m5)+1, s = 1,3,5,...
or
12*(y*3^(s+1)+z*3)+3*3*3^(s+1)+4 =

$12*(y*3^{(s+1)}+z*3+m6)+7$, $s = 2,4,6,...$

Let us note that

$m5 = (3*3*3^{(s+1)}+4 -1)/12 = (3*3^{(s+1)}+1)/4$, when $s = 1,3,5,...,$

$4*z < 3^s$,
$4*z =< 3^s-1$,
$3*z =< 3^{(s+1)}/4-3/4$,
$(3*3^{(s+1)}+1)/4 + 3*z =< 3*3^{(s+1)}/4+ 1/4 + 3^{(s+1)}/4 - 3/4$,
$(3*3^{(s+1)}+1)/4 + 3*z =< 3^{(s+1)} - 1/2$,
$(3*3^{(s+1)}+1)/4+z*3 < 3^{(s+1)}$

and

$m6 = (3*3*3^{(s+1)}+4 -7)/12 = (3*3^{(s+1)}-1)/4$, when $s = 2,4,6,...$

$4*z < 3^s$,
$3*z < 3^{(s+1)}/4$,
$(3*3^{(s+1)}-1)/4 + 3*z < 3*3^{(s+1)}/4 - 1/4 + 3^{(s+1)}/4$,
$(3*3^{(s+1)}-1)/4 + 3*z < 3^{(s+1)} - 1/4$,
$(3*3^{(s+1)}-1)/4+z*3 < 3^{(s+1)}$

Therefore if $v=4*y+3$, $y = 0,1,2,....$, and $t =4*z+0$, then

$12^*(v*3^s+t)+5 \rightarrow 12^*(y*3^{(s+1)}+z*3+(3*3^{(s+1)}+1)/4) +1$,
$s = 1,3,5,...$
or
$12^*(v*3^s+t)+5 \rightarrow 12^*(y*3^{(s+1)}+z*3+(3*3^{(s+1)}-1)/4) +7$,
$s = 2,4,6,...$

this means that the images of natural integers of type 5

$(12*(v*3^s+t)+5)$ in this case will be natural integers of type 1 $(12*(y*3^{(s+1)}+z*3+(3*3^{(s+1)}+1)/4)+1)$, which can be represented as $12*r+1$ and $r = v*3^{(s+1)}+t$, when v = 0,1,2,…. and $t < 3^{(s+1)}$, s = 1,3,5,…

or means that

 the images of natural integers of type 5 $(12*(v*3^s+t)+5)$ in this case will be natural integers of type 7 $(12*(y*3^{(s+1)}+z*3+(3*3^{(s+1)}-1)/4)+7)$, which can be represented as $12*r+7$ and $r = v*3^{(s+1)}+t$, when v = 0,1,2,…. and $t < 3^{(s+1)}$, s = 2,4,6,…

Let $v=4*y+3$, $y = 0,1,2,…$, and $t = 4*z+2$ then

$12*(v*3^s+t)+5 = 12*v*3^s+12*t+5 \rightarrow$
$(3*(12*v*3^s+12*t+5)+1)/4 =$
$(12*((4*y+3)*3^{(s+1)}+(4*z+2)*3)+5*3+1)/4 =$
$(12*((4*y)*3^{(s+1)}+(4*z)*3)+12*3*3^{(s+1)}+12*2*3+5*3+1)/4 =$
$12*(y*3^{(s+1)}+z*3)+3*3*3^{(s+1)}+18+4$

natural integers 3^s can be represented as

in the form $3^s = 12*m1 +3$, when s = 1,3,5,…,
and
in the form $3^s = 12*m2 +9$, when s = 2,4,6,…

 accordingly natural integers $3*3*3^{(s+1)}$ can be represented as

in the form $3*3*3^{(s+1)} = 12*m3 +9$, when s = 1,3,5,…,
and
in the form $3*3*3^{(s+1)} = 12*m4 +3$, when s = 2,4,6,…

accordingly natural integers 3*3*3^(s+1)+18+4 can be represented as

in the form 3*3*3^(s+1)+18+4 = 12*m5 +7, when s = 1,3,5,...,
and
in the form 3*3*3^(s+1)+18+4 = 12*m6 +1, when s = 2,4,6,...

accordingly

12*(y*3^(s+1)+z*3)+3*3*3^(s+1)+18+4 =
12*(y*3^(s+1)+z*3+m5)+7, s = 1,3,5,...
or
12*(y*3^(s+1)+z*3)+3*3*3^(s+1)+18+4 =
12*(y*3^(s+1)+z*3+m6)+1, s = 2,4,6,...

Let us note that

m5 = (3*3*3^(s+1)+18+4 -7)/12 = (3*3^(s+1)+5)/4, when s = 1,3,5,...,

4*z +2< 3^s,
4*z < 3^s-2,
3*z < 3^(s+1)/4-6/4,
(3*3^(s+1)+5)/4 + 3*z < 3*3^(s+1)/4+ 5/4 + 3^(s+1)/4 – 6/4,
(3*3^(s+1)+5)/4 + 3*z < 3^(s+1) – 1/4,
(3*3^(s+1)+5)/4+z*3 < 3^(s+1)

and

m6 = (3*3*3^(s+1)+18+4 -1)/12 = (3*3^(s+1)+7)/4, when s = 2,4,6,...

4*z +2< 3^s,

4*z < 3^s-2,

4*z =< 3^s-2-1,

3*z =< 3^(s+1)/4-9/4,

(3*3^(s+1)+7)/4 + 3*z =< 3*3^(s+1)/4+ 7/4 + 3^(s+1)/4 – 9/4,

(3*3^(s+1)+7)/4 + 3*z =< 3^(s+1) – 1/2,

(3*3^(s+1)+7)/4+z*3 < 3^(s+1)

Therefore if v=4*y+3, y = 0,1,2,…., and t =4*z+2, then

$$12^{*}(v^{*}3^{\wedge}s+t)+5 \rightarrow 12^{*}(y^{*}3^{\wedge}(s+1)+z^{*}3+(3^{*}3^{\wedge}(s+1)+5)/4) +7,$$

s = 1,3,5,…

or

$$12^{*}(v^{*}3^{\wedge}s+t)+5 \rightarrow 12^{*}(y^{*}3^{\wedge}(s+1)+z^{*}3+(3^{*}3^{\wedge}(s+1)+7)/4) +1,$$

s = 2,4,6,…

this means that the images of natural integers of type 5 (12*(v*3^s+t)+5) in this case will be natural integers of type 7 (12*(y*3^(s+1)+z*3+(3*3^(s+1)+5)/4) +7), which can be represented as 12*r+7 and r = v*3^(s+1)+t, when v = 0,1,2,…. and t < 3^(s+1), s = 1,3,5,…

or this means that the images of natural integers of type 5 (12*(v*3^s+t)+5) in this case will be natural integers of type 1 (12*(y*3^(s+1)+z*3+(3*3^(s+1)+7)/4) +1), which can be represented as 12*r+1 and r = v*3^(s+1)+t, when v = 0,1,2,…. and t < 3^(s+1), s = 2,4,6,…

13.1.3.2 Direct horizontal sequences in which the members n, n>1 are natural integers of type 1

Consider direct horizontal sequences that begin in terms of 12*p+q and in which the terms at position n, n>1 are natural integers of type 1

let r = v*3^s+t, where v = 0,1,2,.... and t < 3^s, s>0,

then

12*(v*3^s+t)+1 = 12*v*3^s+12*t+1 =
4*(v*3^(s+1)+3*t)+1

therefore, if the natural integer v is even, then for any odd
natural integer t, the natural integer 4*(v*3^(s+1)+3*t)+1
will be vertical 4*(2*k+1)+1 and therefore this natural
integer will be the last in a direct horizontal sequence that
begins in the natural integer 12*(v*3^s+t)+1 and,
respectively, in the natural integer 12*p+q,

at the same time, if the natural integer v is even, then for
any even natural integer t, the natural integer
4*(v*3^(s+1)+3*t)+1 will be horizontal 4*(2*k)+1 and
therefore this natural integer will not be the last in a direct
horizontal sequence that begins in the natural integer
12*(v*3^s+t)+1 and, respectively, in the natural integer
12*p+q,

further, if the natural integer v is odd, then for any odd
natural integer t the natural integer 4*(v*3^(s+1)+3*t)+1
will be horizontal 4*(2*k)+1 and therefore will not be the
last in a direct horizontal sequence that begins in the natural
integer 12*(v*3^s+t)+1 and, respectively, in the natural
integer 12 * p+q,

at the same time, if the natural integer v is odd, then for any
even natural integer t the natural integer
4*(v*3^(s+1)+3*t)+1 will be vertical 4*(2*k+1)+1 and
therefore the last in a direct horizontal sequence that begins
in the natural integer 12*(v*3^s+t)+1 and, respectively, in

the natural integer 12*p+q.

Therefore, if v = 0,1,2,.... then whatever non-negative integers t are, exactly half of the sequences that begin at 12*p+q and have the element 12*(v*3^s+t) +1 in place of n end at this element and exactly half of the sequences have the next element at n+1

Let's consider what these following elements can be

The following pairs of natural integers v and t are possible, for which the direct horizontal sequences are continued

v=4*y+0, t =4*z+0,
v=4*y+0, t =4*z+2,
v=4*y+2, t =4*z+0,
v=4*y+2, t =4*z+2,
v=4*y+1, t =4*z+1,
v=4*y+1, t =4*z+3,
v=4*y+3, t =4*z+1,
v=4*y+3, t =4*z+3

Let v=4*y+0, y = 0,1,2,…., and t =4*z+0 then

$12^*(v^*3^{\wedge}s+t)+1 = 12^*v^*3^{\wedge}s+12^*t+1 \rightarrow$
$(3^*(12^*v^*3^{\wedge}s+12^*t+1)+1)/4 =$
$(12^*((4^*y+0)^*3^{\wedge}(s+1)+(4^*z+0)^*3)+1^*3+1)/4 =$
$(12^*((4^*y)^*3^{\wedge}(s+1)+(4^*z)^*3)+1^*3+1)/4 =$
$12^*(y^*3^{\wedge}(s+1)+z^*3)+1$

 Let us note that

4*z+0 < 3^s,
z < 3^s,
3*z < 3^(s+1)

Therefore if v=4*y+0, y = 0,1,2,…., and t =4*z+0, then

$$12^{*}(v^{*}3{\wedge}s+t)+1 \rightarrow 12^{*}(y^{*}3{\wedge}(s+1)+z^{*}3)+1$$

this means that the images of natural integers of type 1 12*(v*3^s+t)+1 in this case will be natural integers of type 1 (12*(y*3^(s+1)+z*3)+1), which can be represented as 12*r+1 and r = v*3^(s+1)+t, when v = 0,1,2,…. and t < 3^(s+1), s = 1,2,3,4,…

Let v=4*y+0, y = 0,1,2,…., and t =4*z+2 then

$$12^{*}(v^{*}3{\wedge}s+t)+1 = 12^{*}v^{*}3{\wedge}s+12^{*}t+1 \rightarrow$$
$$(3{\wedge}(12^{*}v^{*}3{\wedge}s+12^{*}t+1)+1)/4 =$$
$$(12^{*}((4^{*}y+0)^{*}3{\wedge}(s+1)+(4^{*}z+2)^{*}3)+1^{*}3+1)/4 =$$
$$(12^{*}((4^{*}y+0)^{*}3{\wedge}(s+1)+(4^{*}z)^{*}3))+12^{*}2^{*}3+1^{*}3+1)/4 =$$
$$12^{*}(y^{*}3{\wedge}(s+1)+z^{*}3)+19 =$$
$$12^{*}(y^{*}3{\wedge}(s+1)+(z^{*}3+1))+7$$

Let us note that

$$4^{*}z+2 < 3{\wedge}s,$$
$$z < 3{\wedge}s/4 - 1/2,$$
$$3^{*}z < 3{\wedge}(s+1)/4 - 3/2,$$
$$3^{*}z + 1 < 3{\wedge}(s+1)/4 - 3/2 + 1,$$
$$3^{*}z+1 < 3{\wedge}(s+1)$$

Therefore if v=4*y+0, y = 0,1,2,…., and t =4*z+2, then

$$12^{*}(v^{*}3{\wedge}s+t)+1 \rightarrow 12^{*}(y^{*}3{\wedge}(s+1)+(z^{*}3+1))+7$$

this means that the images of natural integers of type 1

$(12*(v*3^\wedge s+t)+1)$ in this case will be natural integers of type 7 $(12*(y*3^\wedge(s+1)+(z*3+1))+7)$, which can be represented as $12*r+7$ and $r = v*3^\wedge(s+1)+t$, when $v = 0,1,2,....$ and $t < 3^\wedge(s+1)$, $s = 1,2,3,4,...$

Let $v=4*y+2$, $y = 0,1,2,....$, and $t =4*z+0$ then

$12^*(v*3^\wedge s+t)+1 = 12^*v*3^\wedge s+12^*t+1 \rightarrow$
$(3^*(12^*v*3^\wedge s+12^*t+1)+1)/4 =$
$(12*((4*y+2)*3^\wedge(s+1)+(4*z+0)*3)+1*3+1)/4 =$
$(12*((4*y)*3^\wedge(s+1)+(4*z)*3)+12*2*3^\wedge(s+1)+1*3+1)/4 =$
$12*(y*3^\wedge(s+1)+z*3)+3*2*3^\wedge(s+1)+1$

natural integers $3^\wedge s$ can be represented as

in the form $3^\wedge s = 12*m1 +3$, when $s = 1,3,5,...$,
and
in the form $3^\wedge s = 12*m2 +9$, when $s = 2,4,6,...$

Accordingly natural integers $3*2*3^\wedge(s+1)$ can be represented as

in the form $3*2*3^\wedge(s+1) = 12*m3 +6$, when $s = 1,3,5,...$,
and
in the form $3*2*3^\wedge(s+1) = 12*m4 +6$, when $s = 2,4,6,...$

Accordingly natural integers $3*2*3^\wedge(s+1)+1$ can be represented as

in the form $3*2*3^\wedge(s+1) +1= 12*m5 +7$, when $s = 1,3,5,...$,
and
in the form $3*2*3^\wedge(s+1) +1= 12*m6 +7$, when $s = 2,4,6,...$

Accordingly

$12*(y*3^{(s+1)}+z*3)+3*2*3^{(s+1)}+1 =$
$12*(y*3^{(s+1)}+z*3+m5)+7, s = 1,3,5,...$

or

$12*(y*3^{(s+1)}+z*3))+3*2^3(s+1)+1 =$
$12*(y*3^{(s+1)}+z*3+m6)+7, s = 2,4,6,...$

Let us note that

$m5 = (3*2*3^{(s+1)} - 6)/12 = (3^{(s+1)} - 1)/2$

$4*z+0 < 3^s,$
$z < 3^s/4,$
$3*z < 3^{(s+1)}/4,$
$(3^{(s+1)}-1)/2 < 3^{(s+1)}/2,$
$z*3+(3^{(s+1)}-1)/2 < 3^{(s+1)}/4 + 3^{(s+1)}/2,$
$z*3+(3^{(s+1)}-1)/2 < 3^{(s+1)}$

and

$m6 = (3*2*3^{(s+1)} - 6)/12 = (3^{(s+1)} - 1)/2$

$4*z+0 < 3^s,$
$z < 3^s/4,$
$3*z < 3^{(s+1)}/4,$
$(3^{(s+1)}-1)/2 < 3^{(s+1)}/2,$
$z*3+(3^{(s+1)}-1)/2 < 3^{(s+1)}/4 + 3^{(s+1)}/2,$
$z*3+(3^{(s+1)}-1)/2 < 3^{(s+1)}$

Therefore if $v=4*y+2$, $y = 0,1,2,....$, and $t =4*z+0$, then

$12^{*}(v*3^{\wedge}s+t)+1 \rightarrow 12^{*}(y*3^{\wedge}(s+1)+z*3+(3^{\wedge}(s+1) - 1)/2)+7$

this means that the images of natural integers of type 1
$(12*(v*3^{\wedge}s+t)+1)$ in this case will be natural integers of

type 7 $(12*(y*3^{(s+1)}+z*3+(3^{(s+1)} - 1)/2)+7)$, which can be represented as $12*r+7$ and $r = v*3^{(s+1)}+t$, when v $= 0,1,2,\ldots.$ and $t < 3^{(s+1)}$, $s = 1,2,3,4,\ldots$

Let $v=4*y+2$, $y = 0,1,2,\ldots.$, and $t =4*z+2$ then

$12*(v*3^s+t)+1 = 12*v*3^s+12*t+1 \rightarrow$
$(3^{(12*v*3^s+12*t+1)}+1)/4 =$
$(12*((4*y+2)*3^{(s+1)}+(4*z+2)*3)+1*3+1)/4 =$
$(12*((4*y)*3^{(s+1)}+(4*z)*3)+12*2*3^{(s+1)}+12*2*3+1*3$
$+1)/4 =$
$12*(y*3^{(s+1)}+z*3)+3*2*3^{(s+1)}+19$

natural integers 3^s can be represented as

in the form $3^s = 12*m1 +3$, when $s = 1,3,5,\ldots$,
and
in the form $3^s = 12*m2 +9$, when $s = 2,4,6,\ldots$

Accordingly natural integers $3*2*3^{(s+1)}$ can be represented as

in the form $3*2*3^{(s+1)} = 12*m3 +6$, when $s = 1,3,5,\ldots$,
and
in the form $3*2*3^{(s+1)} = 12*m4 +6$, when $s = 2,4,6,\ldots$

Accordingly natural integers $3*2*3^{(s+1)}+19$ can be represented as

in the form $3*2*3^{(s+1)} +19= 12*m5 +1$, when $s =$ $1,3,5,\ldots$,
and
in the form $3*2*3^{(s+1)} +19= 12*m6 +1$, when $s =$ $2,4,6,\ldots$

Accordingly

$$12*(y*3^{\wedge}(s+1)+z*3)+3*2*3^{\wedge}(s+1)+19 =$$
$$12*(y*3^{\wedge}(s+1)+z*3+m5)+1, \ s = 1,3,5,\ldots$$

or

$$12*(y*3^{\wedge}(s+1)+z*3))+3*2^{\wedge}3(s+1)+19 =$$
$$12*(y*3^{\wedge}(s+1)+z*3+m6)+1, \ s = 2,4,6,\ldots$$

Let us note that

$$m5 = (3*2*3^{\wedge}(s+1) + 18)/12 = (3^{\wedge}(s+1) + 3)/2$$

$$4*z+2 < 3^{\wedge}s,$$
$$4*z < 3^{\wedge}s -2,$$
$$z < 3^{\wedge}s/4 - 1/2,$$
$$3*z < 3^{\wedge}(s+1)/4 - 3/2,$$
$$z*3+(3^{\wedge}(s+1)+3)/2 < 3^{\wedge}(s+1)/4 - 3/2 + 3^{\wedge}(s+1)/2 +3/2,$$
$$z*3+(3^{\wedge}(s+1)+3)/2 < 3^{\wedge}(s+1)/4 + 3^{\wedge}(s+1)/2,$$
$$z*3+(3^{\wedge}(s+1)+3)/2 < 3^{\wedge}(s+1)$$

$$m6 = (3*2*3^{\wedge}(s+1) + 18)/12 = (3^{\wedge}(s+1) + 3)/2$$

$$4*z+2 < 3^{\wedge}s,$$
$$4*z < 3^{\wedge}s -2,$$
$$z < 3^{\wedge}s/4 - 1/2,$$
$$3*z < 3^{\wedge}(s+1)/4 - 3/2,$$
$$z*3+(3^{\wedge}(s+1)+3)/2 < 3^{\wedge}(s+1)/4 - 3/2 + 3^{\wedge}(s+1)/2 +3/2,$$
$$z*3+(3^{\wedge}(s+1)+3)/2 < 3^{\wedge}(s+1)/4 + 3^{\wedge}(s+1)/2,$$
$$z*3+(3^{\wedge}(s+1)+3)/2 < 3^{\wedge}(s+1)$$

Therefore if $v=4*y+2$, $y = 0,1,2,\ldots$, and $t =4*z+2$, then

$$12^{*}(v*3^{\wedge}s+t)+1 \rightarrow 12^{*}(y*3^{\wedge}(s+1)+z*3+(3^{\wedge}(s+1) +3)/2)+7$$

this means that the images of natural integers of type 1 $(12*(v*3^s+t)+1)$ in this case will be natural integers of type 7 $(12*(y*3^{(s+1)}+z*3+(3^{(s+1)} + 3)/2)+7)$, which can be represented as $12*r+7$ and $r = v*3^{(s+1)}+t$, when $v = 0,1,2,....$ and $t < 3^{(s+1)}$, $s = 1,2,3,4,...$

Let $v=4*y+1$, $y = 0,1,2,....$, and $t =4*z+1$ then

$12*(v*3^s+t)+1 = 12*v*3^s+12*t+1 \rightarrow$
$(3*(12*v*3^s+12*t+1)+1)/4 =$
$(12*((4*y+1)*3^{(s+1)}+(4*z+1)*3)+1*3+1)/4 =$
$(12*((4*y)*3^{(s+1)}+(4*z)*3)+12*3^{(s+1)}+12*3+1*3+1)/4$
$=$

$12*(y*3^{(s+1)}+z*3)+3*3^{(s+1)}+9+1 =$

natural integers 3^s can be represented as

in the form $3^s = 12*m1 +3$, when $s = 1,3,5,...,$
and
in the form $3^s = 12*m2 +9$, when $s = 2,4,6,...$

Accordingly natural integers $3*3^{(s+1)}$ can be represented as

in the form $3*3^{(s+1)} = 12*m3 +3$, when $s = 1,3,5,...,$
and
in the form $3*3^{(s+1)} = 12*m4 +9$, when $s = 2,4,6,...$

Accordingly natural integers $3*3^{(s+1)}+10$ can be represented as

in the form $3*3^{(s+1)}+10 = 12*m5 +1$, when $s = 1,3,5,...,$
and
in the form $3*3^{(s+1)}+10 = 12*m6 +7$, when $s = 2,4,6,...$

70

Accordingly natural integers
12*(y*3^(s+1)+z*3)+3*3^(s+1)+10 can be represented as

in the form 12*(y*3^(s+1)+z*3+m5) +1, when s = 1,3,5,...,
and
in the form 12*(y*3^(s+1)+z*3+m6) +7, when s = 2,4,6,...

Let us note that

m5 = (3*3^(s+1)+10 -1)/12 = (3^(s+1)+3)/4, when s = 1,3,5,...,

4*z+1< 3^s,
z< 3^s/4-1/4,
3*z< 3^(s+1)/4-3/4,
(3^(s+1)+3)/4+3*z < 3^(s+1)/4 + ¾ + 3^(s+1)/4-3/4,
(3^(s+1)+3)/4+3*z < 3^(s+1)/2,
(3^(s+1)+3)/4+3*z < 3^(s+1)

or

m6 = (3*3^(s+1)+10 -7)/12 = (3^(s+1)+1)/4, when s = 2,4,6,...

4*z+1< 3^s,
z< 3^s/4-1/4,
3*z< 3^(s+1)/4-3/4,
(3^(s+1)+1)/4+3*z < 3^(s+1)/4 +1/4 + 3^(s+1)/4-3/4,
(3^(s+1)+1)/4+3*z < 3^(s+1)/2 -1/2,
(3^(s+1)+1)/4+3*z < 3^(s+1)

Therefore if v=4*y+1, y = 0,1,2,...., and t =4*z+1, then

$12^*(v^*3\wedge s+t)+1 \rightarrow 12^*(y^*3\wedge(s+1)+(3\wedge(s+1)+3)/4+z^*3)+1$, s = 1,3,5,…

or

$12^*(v^*3\wedge s+t)+1 \rightarrow 12^*(y^*3\wedge(s+1)+(3\wedge(s+1)+1)/4+z^*3)+7$, s = 2,4,6…

this means that the images of natural integers of type 1 $(12*(v*3\wedge s+t)+1)$ in this case there will be natural integers of type 1 $(12*(y*3\wedge(s+1)+(3\wedge(s+1)+3)/4+z*3)+1)$, which can be represented as $12*r+1$ and $r = v*3\wedge(s+1)+t$, when v = 0,1,2,…. and $t < 3\wedge(s+1)$, s = 1,3,5,…

or

there will be natural integers of type 7 $(12*(y*3\wedge(s+1)+(3\wedge(s+1)+1)/4+z*3)+7)$ which can be represented as $12*r+7$ and $r = v*3\wedge(s+1)+t$, when v = 0,1,2,…. and $t < 3\wedge(s+1)$, s = 2,4,6…

Let $v=4*y+1$, y = 0,1,2,…., and $t =4*z+3$ then

$12^*(v^*3\wedge s+t)+1 = 12^*v^*3\wedge s+12^*t+1 \rightarrow$
$(3^*(12^*v^*3\wedge s+12^*t+1)+1)/4 =$
$(12*((4*y+1)*3\wedge(s+1)+(4*z+3)*3)+1*3+1)/4 =$
$(12*((4*y)*3\wedge(s+1)+(4*z)*3)+12*3\wedge(s+1)+12*3*3+1*3+1)/4 =$
$12*(y*3\wedge(s+1)+z*3)+3*3\wedge(s+1)+27+1 =$

natural integers 3^s can be represented as

in the form $3\wedge s = 12*m1 +3$, when s = 1,3,5,…,
and
in the form $3\wedge s = 12*m2 +9$, when s = 2,4,6,…

Accordingly natural integers $3*3^{(s+1)}$ can be represented as

in the form $3*3^{(s+1)} = 12*m3 +3$, when $s = 1,3,5,...$, and
in the form $3*3^{(s+1)} = 12*m4 +9$, when $s = 2,4,6,...$

Accordingly natural integers $3*3^{(s+1)}+27+1$ can be represented as

in the form $3*3^{(s+1)}+27+1 = 12*m5 +7$, when $s = 1,3,5,...$, and
in the form $3*3^{(s+1)}+27+1 = 12*m6 +1$, when $s = 2,4,6,...$

Accordingly natural integers
$12*(y*3^{(s+1)}+z*3)+3*3^{(s+1)}+27+1$ can be represented as

in the form $12*(y*3^{(s+1)}+z*3+m5) +7$, when $s = 1,3,5,...$, and
in the form $12*(y*3^{(s+1)}+z*3+m6) +1$, when $s = 2,4,6,...$

Let us note that

$m5 = (3*3^{(s+1)}+27+1 -7)/12 = (3^{(s+1)}+7)/4$, when $s = 1,3,5,...$,

$4*z+3< 3^s$,
$z< 3^s/4-3/4$,
$3*z< 3^{(s+1)}/4-9/4$,
$(3^{(s+1)}+7)/4 + 3*z < 3^{(s+1)}/4+7/4 + 3^{(s+1)}/4-9/4$,
$(3^{(s+1)}+7)/4 + 3*z < 3^{(s+1)}/4 + 3^{(s+1)}/4 – 1/2$,
$(3^{(s+1)}+7)/4 + 3*z < 3^{(s+1)}/4 + 3^{(s+1)}/4$,

73

$(3^{\wedge}(s+1)+7)/4 + 3*z < 3^{\wedge}(s+1)$

or

m6 = $(3*3^{\wedge}(s+1)+27+1 -1)/12 = (3^{\wedge}(s+1)+9)/4$, when s = 2,4,6,…

$4*z+3< 3^{\wedge}s,$
$z< 3^{\wedge}s/4-3/4,$
$3*z< 3^{\wedge}(s+1)/4-9/4,$
$(3^{\wedge}(s+1)+9)/4 + 3*z < 3^{\wedge}(s+1)/4+9/4 + 3^{\wedge}(s+1)/4-9/4,$
$(3^{\wedge}(s+1)+9)/4 + 3*z < 3^{\wedge}(s+1)/4 + 3^{\wedge}(s+1)/4,$
$(3^{\wedge}(s+1)+9)/4 + 3*z < 3^{\wedge}(s+1)$

Therefore if v=4*y+1, y = 0,1,2,…., and t =4*z+3, then

$12^{\star}(v^{\star}3^{\wedge}s+t)+1 \rightarrow 12^{\star}(y^{\star}3^{\wedge}(s+1)+(3^{\wedge}(s+1)+7)/4+z^{\star}3)+7$, s = 1,3,5,…
or
$12^{\star}(v^{\star}3^{\wedge}s+t)+1 \rightarrow 12^{\star}(y^{\star}3^{\wedge}(s+1)+(3^{\wedge}(s+1)+9)/4+z^{\star}3)+1$, s = 2,4,6...

this means that

 the images of natural integers 12*(v*3^s+t)+1 in this case there will be natural integers of type 7
$(12*(y*3^{\wedge}(s+1)+(3^{\wedge}(s+1)+7)/4+z*3)+7)$, which can be represented as 12*r+7 and r = v*3^(s+1)+t, when v = 0,1,2,…. and t < 3^(s+1), s = 1,3,5,7,…

or

there will be natural integers of type 1
$(12*(y*3^{\wedge}(s+1)+(3^{\wedge}(s+1)+9)/4+z*3)+1)$, which can be represented as 12*r+1 and r = v*3^(s+1)+t, when v =

74

0,1,2,…. and $t < 3^{\wedge}(s+1)$, $s = 2,4,6$…

Let $v=4*y+3$, $y = 0,1,2,….$, and $t = 4*z+1$ then

$12^{*}(v^{*}3^{\wedge}s+t)+1 = 12^{*}v^{*}3^{\wedge}s+12^{*}t+1 \rightarrow$
$(3^{*}(12^{*}v^{*}3^{\wedge}s+12^{*}t+1)+1)/4 =$
$(12^{*}((4^{*}y+3)^{*}3^{\wedge}(s+1)+(4^{*}z+1)^{*}3)+1^{*}3+1)/4 =$
$(12^{*}((4^{*}y)^{*}3^{\wedge}(s+1)+(4^{*}z)^{*}3)+12^{*}3^{*}3^{\wedge}(s+1)+12^{*}3+1^{*}3+1$
$)/4 =$
$12^{*}(y^{*}3^{\wedge}(s+1)+z^{*}3)+3^{*}3^{*}3^{\wedge}(s+1)+9+1 =$

natural integers $3^{\wedge}s$ can be represented as

in the form $3^{\wedge}s = 12^{*}m1 +3$, when $s = 1,3,5,…$,
and
in the form $3^{\wedge}s = 12^{*}m2 +9$, when $s = 2,4,6,…$

Accordingly natural integers $3^{*}3^{*}3^{\wedge}(s+1)$ can be represented as

in the form $3^{*}3^{*}3^{\wedge}(s+1) = 12^{*}m3 +9$, when $s = 1,3,5,…$,
and
in the form $3^{*}3^{*}3^{\wedge}(s+1) = 12^{*}m4 +3$, when $s = 2,4,6,…$

Accordingly natural integers $3^{*}3^{*}3^{\wedge}(s+1)+10$ can be represented as

in the form $3^{*}3^{*}3^{\wedge}(s+1)+10 = 12^{*}m5 +7$, when $s = 1,3,5,…$,
and
in the form $3^{*}3^{*}3^{\wedge}(s+1)+10 = 12^{*}m6 +1$, when $s = 2,4,6,…$

Accordingly natural integers
$12^{*}(y^{*}3^{\wedge}(s+1)+z^{*}3)+3^{*}3^{*}3^{\wedge}(s+1)+10$ can be represented

as

in the form 12*(y*3^(s+1)+z*3+m5) +7, when s = 1,3,5,...,
and
in the form 12*(y*3^(s+1)+z*3+m6) +1, when s = 2,4,6,...

Let us note that

m5 = (3*3*3^(s+1)+10 -7)/12 = (3*3^(s+1)+1)/4, when s = 1,3,5,...,

4*z+1< 3^s,
z< 3^s/4-1/4,
3*z< 3^(s+1)/4-3/4,
3*z + (3*3^(s+1)+1)/4 < 3^(s+1)/4-3/4 + 3^(s+1)*3/4+1/4,
3*z + (3*3^(s+1)+1)/4 < 3^(s+1)/4 + 3^(s+1)*3/4 – 1/2,
3*z + (3*3^(s+1)+1)/4 < 3^(s+1)/4 + 3^(s+1)*3/4,
3*z + (3*3^(s+1)+1)/4 < 3^(s+1)

or

m6 = (3*3*3^(s+1)+10-1)/12 = (3*3^(s+1)+3)/4, when s = 2,4,6,...

4*z+1< 3^s,
z< 3^s/4-1/4,
3*z< 3^(s+1)/4-3/4,
3*z + (3*3^(s+1)+3)/4 < 3^(s+1)/4-3/4 + 3^(s+1)*3/4+3/4,
3*z + (3*3^(s+1)+3)/4 < 3^(s+1)/4 + 3^(s+1)*3/4,
3*z + (3*3^(s+1)+3)/4 < 3^(s+1)

Therefore if v=4*y+3, y = 0,1,2,...., and t =4*z+1, then

12*(v*3^s+t)+1 → 12*(y*3^(s+1)+z*3+(3*3^(s+1)+1)/4) +7,
s = 1,3,5,...
or

76

$12*(v*3^s+t)+1 \rightarrow 12*(y*3^{(s+1)}+z*3+ (3*3^{(s+1)}+3)/4) +1$,
$s = 2,4,6...$

this means that

the images of natural integers $12*(v*3^s+t)+1$ in this case there will be natural integers of type 7 $(12*(y*3^{(s+1)}+z*3+(3*3^{(s+1)}+1)/4) +7)$, which can be represented as $12*r+7$ and $r = v*3^{(s+1)}+t$, when $v = 0,1,2,....$ and $t < 3^{(s+1)}$, $s = 1,3,5,7,...$

or

there will be natural integers of type 1 $(12*(y*3^{(s+1)}+z*3+ (3*3^{(s+1)}+3)/4) +1)$, which can be represented as $12*r+1$ and $r = v*3^{(s+1)}+t$, when $v = 0,1,2,....$ and $t < 3^{(s+1)}$, $s = 2,4,6...$

Let $v=4*y+3$, $y = 0,1,2,....$, and $t =4*z+3$ then

$12*(v*3^s+t)+1 = 12*v*3^s+12*t+1 \rightarrow$
$(3*(12*v*3^s+12*t+1)+1)/4 =$
$(12*((4*y+3)*3^{(s+1)}+(4*z+3)*3)+1*3+1)/4 =$
$(12*((4*y)*3^{(s+1)}+(4*z)*3)+12*3*3^{(s+1)}+12*3*3+1*3 +1)/4 =$
$12*(y*3^{(s+1)}+z*3)+3*3*3^{(s+1)}+27+1 =$

natural integers 3^s can be represented as

in the form $3^s = 12*m1 +3$, when $s = 1,3,5,...$,
and
in the form $3^s = 12*m2 +9$, when $s = 2,4,6,...$

Accordingly natural integers $3*3*3^{(s+1)}$ can be

represented as

in the form $3*3*3^{(s+1)} = 12*m3 +9$, when $s = 1,3,5,\ldots$,
and
in the form $3*3*3^{(s+1)} = 12*m4 +3$, when $s = 2,4,6,\ldots$

Accordingly natural integers $3*3*3^{(s+1)}+27+1$ can be represented as

in the form $3*3*3^{(s+1)}+27+1 = 12*m5 +1$, when $s = 1,3,5,\ldots$,
and
in the form $3*3*3^{(s+1)}+27+1 = 12*m6 +7$, when $s = 2,4,6,\ldots$

Accordingly natural integers
$12*(y*3^{(s+1)}+z*3)+3*3*3^{(s+1)}+27+1$ can be represented as

in the form $12*(y*3^{(s+1)}+z*3+m5) +1$, when $s = 1,3,5,\ldots$,
and
in the form $12*(y*3^{(s+1)}+z*3+m6) +7$, when $s = 2,4,6,\ldots$

Let us note that

$m5 = (3*3*3^{(s+1)}+27+1-1)/12 = (3*3^{(s+1)}+9)/4$, when $s = 1,3,5,\ldots$,

$4*z+3 < 3^s$,
$z < 3^s/4-3/4$,
$3*z < 3^{(s+1)}/4-9/4$,
$3*z + (3*3^{(s+1)}+9)/4 < 3^{(s+1)}/4-9/4 + 3^{(s+1)}*3/4+9/4$,
$3*z + (3*3^{(s+1)}+9)/4 < 3^{(s+1)}/4 + 3^{(s+1)}*3/4$,
$3*z + (3*3^{(s+1)}+9)/4 < 3^{(s+1)}$

or

m6 = (3*3*3^(s+1)+27+1-7)/12 = (3*3^(s+1)+7)/4 , when s = 2,4,6,…

4*z+3< 3^s,
z< 3^s/4-3/4,
3*z< 3^(s+1)/4-9/4,
3*z + (3*3^(s+1)+7)/4 < 3^(s+1)/4 – 9/4 + 3^(s+1)*3/4 + 7/4,
3*z + (3*3^(s+1)+7)/4 < 3^(s+1)/4 + 3^(s+1)*3/4 – 1/2,
3*z + (3*3^(s+1)+7)/4 < 3^(s+1) – 1/2,
3*z + (3*3^(s+1)+7)/4 < 3^(s+1)

Therefore if v=4*y+3, y = 0,1,2,…., and t =4*z+3, then

12*(v*3^s+t)+1 → 12*(y*3^(s+1)+(3*3^(s+1)+9)/4)+z*3)+1, s = 1,3,5,…

or

12*(v*3^s+t)+1 → 12*(y*3^(s+1)+(3*3^(s+1)+7)/4)+z*3)+7, s = 2,4,6...

this means that

 the images of natural integers 12*(v*3^s+t)+1 in this case

there will be natural integers of type 1 (12*(y*3^(s+1)+3*3^(s+1)+9)/4+z*3)+1), which can be represented as 12*r+1 and r = v*3^(s+1)+t, when v = 0,1,2,…. and t < 3^(s+1), s = 1,3,5,7,…

or

there will be natural integers of type 7 (12*(y*3^(s+1)+3*3^(s+1)+7)/4+z*3)+7), which can be represented as 12*r+7 and r = v*3^(s+1)+t, when v = 0,1,2,…. and t < 3^(s+1), s = 2,4,6...

13.1.3.3 Direct horizontal sequences in which the members n, n>1 are natural integers of type 7

Consider direct horizontal sequences that begin in terms of 12*p+q and in which the terms at position n, n>1 are natural integers of type 7

$$12^{*}p+q \rightarrow .. \rightarrow 12^{*}r+7$$

let r = v*3^s+t, where v = 0,1,2,…. and t < 3^s, s>0,

The natural integer 12*r+7 is always horizontal, i.e. cannot be represented as 4*q+1, where q is a natural odd number (see section 11)

Consequently, sequences that start at 12*p+q and have a term 12*(v*3^s+t)+7 in place n always have term on the next place of n+1

Consider what these following members might be

The following pairs of numbers v and t are possible

v=2*y+0, t =2*z+0,
v=2*y+0, t =2*z+1,
v=2*y+1, t =2*z+0,
v=2*y+1, t =2*z+1,

Let v=2*y+0, y = 0,1,2,….,andt =2*z+0 then

80

$12*(v*3^s+t)+7 = 12*v*3^s+12*t+7 \rightarrow$
$(3*(12*v*3^s+12*t+7)+1)/2 =$
$(12*((2*y+0)*3^{\wedge}(s+1)+(2*z+0)*3)+7*3+1)/2 =$
$(12*((2*y)*3^{\wedge}(s+1)+(2*z)*3)+7*3+1)/2 =$
$12*(y*3^{\wedge}(s+1)+z*3)+11$

Let us note that

$2*z+0< 3^s,$
$z< 3^{\wedge}s/2,$
$3*z< 3^{\wedge}(s+1)/2,$
$3*z< 3^{\wedge}(s+1),$

Therefore if v=2*y+0, y = 0,1,2,...., and t =2*z+0, then

$12*(v*3^s+t)+7 \rightarrow 12*(y*3^{\wedge}(s+1)+z*3)+11$

this means that the images of natural integers of type 7 $(12*(v*3^s+t)+7)$ in this case will be natural integers of type 11, $12*(y*3^{\wedge}(s+1)+z*3)+11$, which can be represented as $12*r+11$ and $r = v*3^{\wedge}(s+1)+t$, when v = 0,1,2,.... and $t < 3^{\wedge}(s+1)$, s = 1,2,3,4,...

Let v=2*y+0, y = 0,1,2,...., and t =2*z+1 then

$12*(v*3^s+t)+7 = 12*v*3^s+12*t+7 \rightarrow$
$(3*(12*v*3^s+12*t+7)+1)/2 =$
$(12*((2*y+0)*3^{\wedge}(s+1)+(2*z+1)*3)+7*3+1)/2 =$
$(12*((2*y)*3^{\wedge}(s+1)+(2*z)*3)+12*3+7*3+1)/2 =$
$12*(y*3^{\wedge}(s+1)+z*3)+29 = 12*(y*3^{\wedge}(s+1)+z*3+2)+5$

Let us note that

$2*z+1< 3^s,$

$z < 3^s/2 - 1/2$,

$3*z < 3^{(s+1)}/2 - 3/2$,

$3*z < 3^{(s+1)}/2$,

$2 < 3^{(s+1)}/2$, s =1,2,3,…

$3*z+2 < 3^{(s+1)}$,

Therefore if $v=2*y+0$, $y = 0,1,2,…$, and $t =2*z+1$, then

$$12*(v*3^s+t)+7 \rightarrow 12*(y*3^{(s+1)}+z*3+2)+5$$

this means that the images of natural integers of type 7 $(12*(v*3^s+t)+7)$ in this case will be natural integers of type 5, $12*(y*3^{(s+1)}+z*3+2)+5$, which can be represented as $12*r+5$ and $r = v*3^{(s+1)}+t$, when $v = 0,1,2,…$ and $t < 3^{(s+1)}$, $s = 1,2,3,4,…$

Let $v=2*y+1$, $y = 0,1,2,…$, and $t =2*z+0$ then

$12*(v*3^s+t)+7 = 12*v*3^s+12*t+7 \rightarrow$
$(3*(12*v*3^s+12*t+7)+1)/2 =$
$(12*((2*y+1)*3^{(s+1)}+(2*z+0)*3)+7*3+1)/2 =$
$(12*((2*y)*3^{(s+1)}+(2*z)*3)+12*3^{(s+1)}+7*3+1)/2 =$
$12*(y*3^{(s+1)}+z*3)+6*3^{(s+1)}+11$

Natural integers 3^s can be represented as

in the form $3^s = 12*m1 +3$, when $s = 1,3,5,…$,
and
in the form $3^s = 12*m2 +9$, when $s = 2,4,6,…$

Accordingly natural integers $6*3^{(s+1)}$ can be represented as

in the form $6*3^{(s+1)} = 12*m3 +6$, when $s = 1,3,5,…$,

and
in the form $6*3^{(s+1)} = 12*m4 +6$, when s = 2,4,6,…

Accordingly natural integers $6*3^{(s+1)}+11$ can be represented as

in the form $6*3^{(s+1)}+11 = 12*m5 +5$, when s = 1,3,5,…,
and
in the form $6*3^{(s+1)}+11 = 12*m6 +5$, when s = 2,4,6,…

Accordingly natural integers
$12*(y*3^{(s+1)}+z*3)+6*3^{(s+1)}+11$ can be represented as

in the form $12*(y*3^{(s+1)}+z*3+m5) +5$, when s = 1,3,5,…,
and
in the form $12*(y*3^{(s+1)}+z*3+m6) +5$, when s = 2,4,6,…

Let us note that

$m5 = (6*3^{(s+1)}+11-5)/12 = (3^{(s+1)}+1)/2$, when s = 1,3,5,…,

$2*z+0< 3^s$,
$2*z =< 3^s – 1$,
$z =< 3^s/2 – 1/2$,
$3*z =< 3^{(s+1)}/2 – 3/2$,
$3*z + (3^{(s+1)}+1)/2 =< 3^{(s+1)}/2 – 3/2 + 3^{(s+1)}/2 +1/2$,
$3*z + (3^{(s+1)}+1)/2 =< 3^{(s+1)}/2 + 3^{(s+1)}/2 - 1$,
$3*z + (3^{(s+1)}+1)/2 < 3^{(s+1)}/2 + 3^{(s+1)}/2$,
$3*z + (3^{(s+1)}+1)/2 < 3^{(s+1)}$

or

$m6 = (6*3^{(s+1)}+11-5)/12 = (3^{(s+1)}+1)/2$, when s = 2,4,6,…,

$2*z+0 < 3\text{^}s,$

$2*z =< 3\text{^}s - 1,$

$z =< 3\text{^}s/2 - 1/2,$

$3*z =< 3\text{^}(s+1)/2 - 3/2,$

$3*z + (3\text{^}(s+1)+1)/2 =< 3\text{^}(s+1)/2 - 3/2 + 3\text{^}(s+1)/2 +1/2,$

$3*z + (3\text{^}(s+1)+1)/2 =< 3\text{^}(s+1)/2 + 3\text{^}(s+1)/2 - 1,$

$3*z + (3\text{^}(s+1)+1)/2 < 3\text{^}(s+1)/2 + 3\text{^}(s+1)/2,$

$3*z + (3\text{^}(s+1)+1)/2 < 3\text{^}(s+1)$

Therefore if $v=2*y+1$, $y = 0,1,2,....$, and $t =2*z+0$, then

$$12*(v*3\text{^}s+t)+7 \rightarrow 12*(y*3\text{^}(s+1)+3*z + (3\text{^}(s+1)+1)/2)+5$$

this means that the images of natural integers of type 7 $(12*(v*3\text{^}s+t)+7)$ in this case will be natural integers of type 5, $12*(y*3\text{^}(s+1)+3*z + (3\text{^}(s+1)+1)/2)+5$, which can be represented as $12*r+5$ and $r = v*3\text{^}(s+1)+t$, when $v = 0,1,2,....$ and $t < 3\text{^}(s+1)$, $s = 1,2,3,4,...$

Let $v=2*y+1$, $y = 0,1,2,....$, and $t =2*z+1$ then

$12*(v*3\text{^}s+t)+7 = 12*v*3\text{^}s+12*t+7 \rightarrow$

$(3*(12*v*3\text{^}s+12*t+7)+1)/2 =$

$(12*((2*y+1)*3\text{^}(s+1)+(2*z+1)*3)+7*3+1)/2 =$

$(12*((2*y)*3\text{^}(s+1)+(2*z)*3)+12*3\text{^}(s+1)+12*3+7*3+1)/2$

$=$

$12*(y*3\text{^}(s+1)+z*3)+6*3\text{^}(s+1)+18+11 =$

$12*(y*3\text{^}(s+1)+z*3)+6*3\text{^}(s+1)+29$

Natural integers $3\text{^}s$ can be represented as

in the form $3\text{^}s = 12*m1 +3$, when $s = 1,3,5,...$,
and
in the form $3\text{^}s = 12*m2 +9$, when $s = 2,4,6,...$

Accordingly natural integers 6*3^(s+1) can be represented as

in the form 6*3^(s+1) = 12*m3 +6, when s = 1,3,5,...,
and
in the form 6*3^(s+1) = 12*m4 +6, when s = 2,4,6,...

Accordingly natural integers 6*3^(s+1)+29 can be represented as

in the form 6*3^(s+1)+29 = 12*m5 +11, when s = 1,3,5,...,
and
in the form 6*3^(s+1)+29 = 12*m6 +11, when s = 2,4,6,...

Accordingly natural integers
12*(y*3^(s+1)+z*3)+6*3^(s+1)+29 can be represented as

in the form 12*(y*3^(s+1)+z*3+m5) +11, when s = 1,3,5,...,
and
in the form 12*(y*3^(s+1)+z*3+m6) +11, when s = 2,4,6,...

Let us note that

m5 = (6*3^(s+1)+29-11)/12 = (3^(s+1)+3)/2, when s = 1,3,5,...,

2*z+1< 3^s,
z< 3^s/2 – 1/2,
3*z< 3^(s+1)/2 – 3/2,
3*z + (3^(s+1)+3)/2 < 3^(s+1)/2 – 3/2 + 3^(s+1)/2 +3/2,
3*z + (3^(s+1)+3)/2 < 3^(s+1)

or

m6 = (6*3^(s+1)+29-11)/12 = (3^(s+1)+3)/2, when s = 2,4,6,…,

$2*z+1< 3^s$,
$z< 3^s/2 – 1/2$,
$3*z< 3^{(s+1)}/2 – 3/2$,
$3*z + (3^{(s+1)}+3)/2 < 3^{(s+1)}/2 – 3/2 + 3^{(s+1)}/2 +3/2$,
$3*z + (3^{(s+1)}+3)/2 < 3^{(s+1)}$

Therefore if v=2*y+1, y = 0,1,2,…., and t =2*z+1, then

$$12^*(v^*3\text{^}s+t)+7 \to 12^*(y^*3\text{^}(s+1)+3^*z + (3\text{^}(s+1)+3)/2)+11$$

this means that the images of natural integers of type 7 (12*(v*3^s+t)+7) in this case will be natural integers of type 11 (12*(y*3^(s+1)+3*z + (3^(s+1)+3)/2)+11), which can be represented as 12*r+11 and r = v*3^(s+1)+t, when v = 0,1,2,…. and t < 3^(s+1), s = 1,2,3,4,…

13.1.3.4 Direct horizontal sequences in which the members n, n>1 are natural integers of type 11

Consider direct horizontal sequences that begin in terms of 12*p+q and in which the terms at position n, n>1 are natural integers of type 11

$$12^*p+q \to .. \to 12^*r+11$$

let r = v*3^s+t, where v = 0,1,2,…. and t < 3^s, s>0,

The natural integer 12*r+11 is always horizontal, i.e. cannot be represented as 4*q+1, where q is a natural odd number (see section 11)

Consequently, sequences that start at 12*p+q and have a
term 12*(v*3^s+t)+11 in place n always have term on the
next place of n+1

Consider what these following members might be

The following pairs of numbers v and t are possible

v=2*y+0, t =2*z+0,
v=2*y+0, t =2*z+1,
v=2*y+1, t =2*z+0,
v=2*y+1, t =2*z+1,

Let v=2*y+0, y = 0,1,2,...., and t =2*z+0 then

$12^*(v^*3\text{^}s+t)+11 = 12^*v^*3\text{^}s+12^*t+11 \rightarrow$
$(3^*(12^*v^*3\text{^}s+12^*t+11)+1)/2 =$
$(12^*((2^*y+0)^*3\text{^}(s+1)+(2^*z+0)^*3)+11^*3+1)/2 =$
$(12^*((2^*y)^*3\text{^}(s+1)+(2^*z)^*3)+11^*3+1)/2 =$
$12^*(y^*3\text{^}(s+1)+z^*3)+17 =$
$12^*(y^*3\text{^}(s+1)+(z^*3+1))+5$

Let us note that

2*z+0< 3^s,
z< 3^s/2,
3*z< 3^(s+1)/2,
1 < 3^(s+1)/2, s =1,2,3,...
3*z+1< 3^(s+1)/2 + 3^(s+1)/2,
3*z+1< 3^(s+1),

Therefore if v=2*y+0, y = 0,1,2,...., and t =2*z+0, then

$12^*(v^*3{\wedge}s+t)+11 \rightarrow 12^*(y^*3{\wedge}(s+1)+(3^*z+1))+5$

this means that the images of natural integers of type 11 $(12^*(v^*3{\wedge}s+t)+11)$ in this case will be natural integers of type 5, $12^*(y^*3{\wedge}(s+1)+3^*z+1)+5$, which can be represented as 12^*r+5 and $r = v^*3{\wedge}(s+1)+t$, when $v = 0,1,2,\ldots$ and $t < 3{\wedge}(s+1)$, $s = 1,2,3,4,\ldots$

Let $v=2^*y+0$, $y = 0,1,2,\ldots$, and $t = 2^*z+1$ then

$12^*(v^*3{\wedge}s+t)+11 = 12^*v^*3{\wedge}s+12^*t+11 \rightarrow$
$(3^*(12^*v^*3{\wedge}s+12^*t+11)+1)/2 =$
$(12^*((2^*y+0)^*3{\wedge}(s+1)+(2^*z+1)^*3)+11^*3+1)/2 =$
$(12^*((2^*y)^*3{\wedge}(s+1)+(2^*z)^*3)+12^*3+11^*3+1)/2 =$
$12^*(y^*3{\wedge}(s+1)+z^*3)+35 =$
$12^*(y^*3{\wedge}(s+1)+(z^*3+2))+11$

Let us note that

$2^*z+1 < 3{\wedge}s,$
$z < 3{\wedge}s/2 - 1/2,$
$3^*z < 3{\wedge}(s+1)/2 - 3/2,$
$2 < 3{\wedge}(s+1)/2, s = 1,2,3,\ldots$
$3^*z+2 < 3{\wedge}(s+1)/2 - 3/2 + 3{\wedge}(s+1)/2$
$3^*z+2 < 3{\wedge}(s+1),$

Therefore if $v=2^*y+0$, $y = 0,1,2,\ldots$, and $t = 2^*z+1$, then

$12^*(v^*3{\wedge}s+t)+11 \rightarrow 12^*(y^*3{\wedge}(s+1)+(3^*z+2))+11$

this means that the images of natural integers of type 11 $(12^*(v^*3{\wedge}s+t)+11)$ in this case will be natural integers of type 11, $12^*(y^*3{\wedge}(s+1)+(z^*3+2))+11$, which can be represented as 12^*r+11 and $r = v^*3{\wedge}(s+1)+t$, when $v =$

$0,1,2,\ldots$ and $t < 3^{\wedge}(s+1)$, $s = 1,2,3,4,\ldots$

Let $v=2*y+1$, $y = 0,1,2,\ldots$, and $t =2*z+0$ then

$12*(v*3^{\wedge}s+t)+11 = 12*v*3^{\wedge}s+12*t+11 \rightarrow$
$(3*(12*v*3^{\wedge}s+12*t+11)+1)/2 =$
$(12*((2*y+1)*3^{\wedge}(s+1)+(2*z+0)*3)+11*3+1)/2 =$
$(12*((2*y)*3^{\wedge}(s+1)+(2*z)*3)+12*3^{\wedge}(s+1)+11*3+1)/2 =$
$12*(y*3^{\wedge}(s+1)+z*3)+6*3^{\wedge}(s+1)+17$

Natural integers $3^{\wedge}s$ can be represented as

in the form $3^{\wedge}s = 12*m1 +3$, when $s = 1,3,5,\ldots$,
and
in the form $3^{\wedge}s = 12*m2 +9$, when $s = 2,4,6,\ldots$

Accordingly natural integers $6*3^{\wedge}(s+1)$ can be represented as

in the form $6*3^{\wedge}(s+1) = 12*m3 +6$, when $s = 1,3,5,\ldots$,
and
in the form $6*3^{\wedge}(s+1) = 12*m4 +6$, when $s = 2,4,6,\ldots$

Accordingly natural integers $6*3^{\wedge}(s+1)+17$ can be represented as

in the form $6*3^{\wedge}(s+1)+17 = 12*m5 +11$, when $s = 1,3,5,\ldots$,
and
in the form $6*3^{\wedge}(s+1)+17 = 12*m6 +11$, when $s = 2,4,6,\ldots$

Accordingly natural integers
$12*(y*3^{\wedge}(s+1)+z*3)+6*3^{\wedge}(s+1)+17$ can be represented as

in the form $12*(y*3^{\wedge}(s+1)+z*3+m5) +11$, when $s = 1,3,5,\ldots$,

and
in the form $12*(y*3^{(s+1)}+z*3+m6) +11$, when s = 2,4,6,…

Let us note that

m5 = $(6*3^{(s+1)}+17-11)/12 = (3^{(s+1)}+1)/2$, when s = 1,3,5,…,

$2*z+0< 3^s$,
$2*z =< 3^s - 1$,
$z =< 3^s/2 – 1/2$,
$3*z =< 3^{(s+1)}/2 – 3/2$,
$3^{(s+1)}/2 +1/2 < 3^{(s+1)}/2 +1$,
$3*z + (3^{(s+1)}+1)/2 < 3^{(s+1)}/2 – 3/2 + 3^{(s+1)}/2 +1$,
$3*z + (3^{(s+1)}+1)/2 < 3^{(s+1)}/2 + 3^{(s+1)}/2 – 1/2$,
$3*z + (3^{(s+1)}+1)/2)< 3^{(s+1)}$

or

m6 = $(6*3^{(s+1)}+17-11)/12 = (3^{(s+1)}+1)/2$, when s = 2,4,6,…

$2*z+0< 3^s$,
$2*z =< 3^s - 1$,
$z =< 3^s/2 – 1/2$,
$3*z =< 3^{(s+1)}/2 – 3/2$,
$3^{(s+1)}/2 +1/2 < 3^{(s+1)}/2 +1$,
$3*z + (3^{(s+1)}+1)/2 < 3^{(s+1)}/2 – 3/2 + 3^{(s+1)}/2 +1$,
$3*z + (3^{(s+1)}+1)/2 < 3^{(s+1)}/2 + 3^{(s+1)}/2 – 1/2$,
$3*z + (3^{(s+1)}+1)/2)< 3^{(s+1)}$

Therefore if v=2*y+1, y = 0,1,2,…., and t =2*z+0, then

$12^*(v^*3^{\wedge}s+t)+11 \rightarrow 12^*(y^*3^{\wedge}(s+1)+3^*z + (3^{\wedge}(s+1)+1)/2)+11$

this means that the images of natural integers of type 11 $(12*(v*3^s+t)+11)$ in this case will be natural integers of type 11 $(12*(y*3^{(s+1)}+3*z + (3^{(s+1)}+1)/2)+11)$, which can be represented as $12*r+11$ and $r = v*3^{(s+1)}+t$, when v = 0,1,2,.... and $t < 3^{(s+1)}$, s = 1,2,3,4,...

Let v=2*y+1, y = 0,1,2,...., and t =2*z+1 then

$12^*(v^*3\text{^}s+t)+11 = 12^*v^*3\text{^}s+12^*t+11 \rightarrow$
$(3^*(12^*v^*3\text{^}s+12^*t+11)+1)/2 =$
$(12*((2*y+1)*3\text{^}(s+1)+(2*z+1)*3)+11*3+1)/2 =$
$(12*((2*y)*3\text{^}(s+1)+(2*z)*3)+12*3\text{^}(s+1)+12*3+11*3+1)/2 =$
$12*(y*3\text{^}(s+1)+z*3)+6*3\text{^}(s+1)+18+17 =$
$12*(y*3\text{^}(s+1)+z*3)+6*3\text{^}(s+1)+35 =$
Natural integers $3\text{^}s$ can be represented as

in the form $3\text{^}s = 12*m1 +3$, when s = 1,3,5,...,
and
in the form $3\text{^}s = 12*m2 +9$, when s = 2,4,6,...

Accordingly natural integers $6*3\text{^}(s+1)$ can be represented as

in the form $6*3\text{^}(s+1) = 12*m3 +6$, when s = 1,3,5,...,
and
in the form $6*3\text{^}(s+1) = 12*m4 +6$, when s = 2,4,6,...

Accordingly natural integers $6*3\text{^}(s+1)+35$ can be represented as

in the form $6*3\text{^}(s+1)+35 = 12*m5 +5$, when s = 1,3,5,...,
and
in the form $6*3\text{^}(s+1)+35 = 12*m6 +5$, when s = 2,4,6,...

91

Accordingly natural integers
12*(y*3^(s+1)+z*3)+6*3^(s+1)+35 can be represented as

in the form 12*(y*3^(s+1)+z*3+m5) +5, when s = 1,3,5,…,
and
in the form 12*(y*3^(s+1)+z*3+m6) +5, when s = 2,4,6,…

Let us note that

m5 = (6*3^(s+1)+35-5)/12 = (3^(s+1)+5)/2, when s = 1,3,5,…,

2*z+1< 3^s,
2*z < 3^s – 1,
2*z =< 3^s – 3,
3*z =< (3^(s+1) – 9)/2 ,
3*z + (3^(s+1)+5)/2=< (3^(s+1) – 9)/2 + 3^(s+1)/2+5/2,
3*z + (3^(s+1)+5)/2=< 3^(s+1)/2 – 9/2 + 3^(s+1)/2+5/2,
3*z + (3^(s+1)+5)/2=< 3^(s+1)- 2,
3*z + (3^(s+1)+5)/2< 3^(s+1)

or

m6 = (6*3^(s+1)+35-5)/12 = (3^(s+1)+5)/2, when s = 2,4,6,…,

2*z+1< 3^s,
2*z < 3^s – 1,
2*z =< 3^s – 3,
3*z =< (3^(s+1) – 9)/2 ,
3*z + (3^(s+1)+5)/2=< (3^(s+1) – 9)/2 + 3^(s+1)/2+5/2,
3*z + (3^(s+1)+5)/2=< 3^(s+1)/2 – 9/2 + 3^(s+1)/2+5/2,
3*z + (3^(s+1)+5)/2=< 3^(s+1)- 2,
3*z + (3^(s+1)+5)/2< 3^(s+1)

Therefore if v=2*y+1, y = 0,1,2,…., and t =2*z+1, then

$$12^*(v*3{\wedge}s{+}t){+}11 \rightarrow 12^*(y*3{\wedge}(s{+}1){+}3^*z + (3{\wedge}(s{+}1){+}5)/2){+}5$$

this means that the images of natural integers of type 11 $(12*(v*3{\wedge}s{+}t){+}11)$ in this case will be natural integers of type 5 $(12*(y*3{\wedge}(s{+}1){+}3*z + (3{\wedge}(s{+}1){+}5)/2){+}5)$, which can be represented as $12*r{+}5$ and $r = v*3{\wedge}(s{+}1){+}t$, when $v = 0,1,2,....$ and $t < 3{\wedge}(s{+}1)$, $s = 1,2,3,4,...$

13.1.3.5 Consequence from sections 13.1.3.1 to 13.1.3.4

It follows from sections 13.1.3.1 to 13.1.3.4 and 12.1 that for any $p >= 0$ if $q = 3,5,7,9,11$, and $p > 0$ if $q =1$, and any $m > 0$ in a set of reverse horizontal sequences

$12 * p{+}q$, $12*p{+}(q{+}2)$, $12 * p{+}(q{+}4)$, $12*p{+}(q{+}6)$, $12*p{+}(q{+}8)$, $12*p{+}(q{+}10),12*(p{+}1){+}q$, $12*(p{+}1){+}(q{+}2)$, $12*(p{+}1){+}(q{+}4)$, $12*(p{+}1){+}(q{+}6)$, $12*(p{+}1){+}(q{+}8)$, $12*(p{+}1){+}(q{+}10),......,12*(p{+}4{\wedge}(n{-}1)*4{\wedge}m{-}1){+}q$, $12*(p{+}4{\wedge}(n{-}1)*4{\wedge}m{-}1){+}(q{+}2)$, $12*(p{+}4{\wedge}(n{-}1)*4{\wedge}m{-}1){+}(q{+}4)$, $12*(p{+}4{\wedge}(n{-}1)*4{\wedge}m{-}1){+}(q{+}6)$, $12*(p{+}4{\wedge}(n{-}1)*4{\wedge}m{-}1){+}(q{+}8)$, $12*(p{+}4{\wedge}(n{-}1)*4{\wedge}m{-}1){+}(q{+}10)$

exactly $(6*4{\wedge}(n{-}1)*4{\wedge}m)*(3/4){\wedge}n$ have length greater than n.

Hence (see section 12.1) the proportion of sequences whose length is greater than n in any set (whose size is a multiple of $6*4{\wedge}(n{-}1)*4{\wedge}n$) of straight horizontal sequences that are arranged in a row is $(3/4){\wedge}n$.

13.2 Rule (2/3)^n for sequences whose lengths are greater than n in a set of reverse horizontal sequences.

Then everywhere in section 13.2 the symbol $\rightarrow$ will denote a reverse horizontal conversion G and s will denote a positive integer

13.2.1 the Rule (2/3)^1 for sequences with lengths that are greater than 1 in the set of reverse horizontal sequences.

Consider reverse horizontal sequences that begin in natural integers of type 3 or type 9

Natural integers of type 3 or type 9 have not the image with the reverse horizontal conversion, (see section 11) therefore, all reverse horizontal sequences that start at these natural integers have a length of 1 and end at these natural integers.

Hence all the reverse horizontal sequences, which start in natural integers $12*p + 3$ and $12*p+9$, p=0,1,2,.. have a length of 1

Consider the reverse horizontal sequences that begin with natural integer with type 7

$12*0 + 7$, $12*1 + 7$, $12*2+ 7$ and so on

all natural integers $12*p+7$, p=0,1,.. always have an image in reverse horizontal conversion,

if $p = 3*k+0$, then $(4*(12*(3*k+0)+7)-1)/3 = 12*(4*k +0) + 9$,

94

if p = 3*k+1, then (4*(12*(3*k+1)+7)-1)/3 = 12*(4*k +2) + 1,

if p = 3*k+2, then (4*(12*(3*k+2)+7)-1)/3 = 12*(4*k +3) + 5,

Hence all the reverse horizontal sequences, which starts in numbers 12*p + 7, p=0,1,2,.. have a length greater than 1

Consider the reverse horizontal sequences that begin with natural integer with type 11

12*0 + 11, 12*1 + 11, 12*2+ 11 and so on

all natural integers 12*p+11, p=0,1,.. always have an image in reverse horizontal conversion,

if p = 3*k+0, then (2*(12*(3*k+0)+11)-1)/3 = 12*(2*k +0) + 7,

if p = 3*k+1, then (2*(12*(3*k+1)+11)-1)/3 = 12*(3*k +1) + 3,

if p = 3*k+2, then (2*(12*(3*k+2)+11)-1)/3 = 12*(3*k +1) + 11,

Hence all the reverse horizontal sequences, which starts in numbers 12*p + 11, p=0,1,2,.. have a length greater than 1

Consider the reverse horizontal sequences that begin with natural integer with type 5

12*0 + 5, 12*1 + 5, 12*2+ 5 and so on

95

all natural integers 12*p+5, p=0,1,.. always have an image in reverse horizontal conversion,

if p = 3*k+0, then (2*(12*(3*k+0)+5)-1)/3 = 12*(2*k +0) + 3,

if p = 3*k+1, then (2*(12*(3*k+1)+5)-1)/3 = 12*(4*k +0) + 11,

if p = 3*k+2, then (2*(12*(3*k+2)+5)-1)/3 = 12*(4*k +1) + 7,

Hence all the reverse horizontal sequences, which starts in numbers 12*p + 5, p=0,1,2,.. have a length greater than 1

Consider the reverse horizontal sequences that begin with natural integer with type 1

12*1 + 1, 12*2 + 1, 12*3+ 1 and so on

all natural integers 12*p+1, p=1,2,.. always have an image in reverse horizontal conversion,

if p = 3*k+0, then (4*(12*(3*k+0)+1)-1)/3 = 12*(2*k +4) + 1,

if p = 3*k+1, then (4*(12*(3*k+1)+1)-1)/3 = 12*(4*k +1) + 5,

if p = 3*k+2, then (4*(12*(3*k+2)+1)-1)/3 = 12*(4*k +2) + 9,

Hence all the reverse horizontal sequences, which starts in numbers 12*p + 1, p=0,1,2,.. have a length greater than 1

From the above reasoning and section 12.2, it follows that for any p > = 0 if q = 3,5,7,9,11, and p > 0 if q =1, and any m > 0 in the set of reverse horizontal sequences

12*p+q, 12*p+(q+2), 12*p+(q+4), 12*p+(q+6),
12*p+(q+8), 12*p+(q+10),
12*(p+1)+q, 12*(p+1)+(q+2), 12*(p+1)+(q+4),
12*(p+1)+(q+6),
12*(p+1)+(q+8),12*(p+1)+(q+10),....,12*(p+3^m-1)+q,
12*(p+3^m-1)+(q+2), 12*(p+3^m-1)+(q+4), 12*(p+3^m-1)+(q+6), 12*(p+3^m-1)+(q+8), 12*(p+3^m-1)+(q+10)

exactly $(6*3^m)*(2/3)^1$ have a length greater than 1,

namely 12 from 18, 24 from 36 and etc.

Hence (see section 12.2) the proportion of reverse horizontal sequences whose length is greater than 1 in any set (whose size is a multiple of 6*3^1) of reverse horizontal sequences that are arranged in a row is $(2/3^1$

13.2.2 Rule (3/4)^2 for sequences with lengths that are greater than 2 in the set of reverse horizontal sequences.

Consider reverse horizontal sequences that begin in natural integers of type 5, where p = 0,3,6,...

$12*0 + 5 \rightarrow 12*0 + 3$,
$12*3 + 5 \rightarrow 12*2 + 3$,
$12*6 + 5 \rightarrow 12*4 + 3$,

Note that all natural integers that stand at position number 2 in the considered reverse horizontal sequences can be

represented as 12*r+3, where r = v*2^s+t, at v = 0,1,2,.....
and t < 2^s, where s = 1.

Hence all sequences that begin in elements of type 5,
12*p+5, at p = 0,3,6,..in position 2 have an element of type
3 element, which has no image in the reverse horizontal
transformation, and hence all those reverse horizontal
sequences are of length 2

Consider reverse horizontal sequences that begin in natural
integers of type 5, where p = 1,4,7,…

$$12^*1 + 5 \rightarrow 12^*1 + 11,$$
$$12^*4 + 5 \rightarrow 12^*3 + 11,$$
$$12^*7 + 5 \rightarrow 12^*5 + 11,$$

Note that all natural integers that stand at position number 2
in the considered reverse horizontal sequences can be
represented as 12*r+11, where r = v*2^s+t, at v = 0,1,2,.....
and t < 2^s, where s = 1.

Hence, in all sequences that begin in elements of type 5,
12*p+5, at p = 1,4,7,.. in position 2 have an element of type
11, which has an image in the reverse horizontal
transformation, and therefore all the reverse horizontal
sequences have a length greater than 2

Consider reverse horizontal sequences that begin in natural
integers of type 5, where p = 2,5,8,…

$$12^*2 + 5 \rightarrow 12^*1 + 7,$$
$$12^*5 + 5 \rightarrow 12^*3 + 7,$$
$$12^*8 + 5 \rightarrow 12^*5 + 7,$$

Note that all natural integers that stand at position number 2 in the considered reverse horizontal sequences can be represented as 12*r+7, where r = v*2^s+t, at v = 0,1,2,..... and t < 2^s, where s = 1.

Hence, in all sequences that begin in elements of type 5, 12*p+5, at p = 2,5,8,.. in position 2 have an element of type 7, which has an image in the reverse horizontal transformation, and therefore all the reverse horizontal sequences have a length greater than 2

Therefore, the proportion of sequences that begin in numbers 12*p+5 and have a length greater than 2 is 2/3 of the set of sequences that begin in natural integer of type 12*p+5 and whose length is greater than 1

Consider reverse horizontal sequences that begin in natural integers of type 7, p = 0,3,6,...

$12^*0 + 7 \rightarrow 12^*0 + 9,$
$12^*3 + 7 \rightarrow 12^*4 + 9,$
$12^*6 + 7 \rightarrow 12^*8 + 9,$

Note that all natural integers that stand at position number 2 in the considered reverse horizontal sequences can be represented as 12*r+9, where r = v*4^s+t, at v = 0,1,2,..... and t < 4^s, where s = 1.

Hence all sequences that begin in elements of type 7, 12*p+7, at p = 0,3,6,.. in position 2 have an element of type 9 element, which has no image in the reverse horizontal transformation, and hence all those reverse horizontal sequences are of length 2

Consider reverse horizontal sequences that begin in natural

integers of type 7, where p = 1,4,7,...

$12^{*}1 + 7 \rightarrow 12^{*}2 + 1,$
$12^{*}4 + 7 \rightarrow 12^{*}6 + 1,$
$12^{*}7 + 7 \rightarrow 12^{*}10 + 1,$

Note that all natural integers that stand at position number 2 in the considered reverse horizontal sequences can be represented as $12*r+1$, where $r = v*4^{\wedge}s+t$, at $v = 0,1,2,.....$ and $t < 4^{\wedge}s$, where $s = 1$.

Hence, in all sequences that begin in elements of type 7, $12*p+7$, at $p = 1,4,7,..$ in position 2 have an element of type 1, which has an image in the reverse horizontal transformation, and therefore all the reverse horizontal sequences have a length greater than 2

Consider reverse horizontal sequences that begin in natural integers of type 7, where p = 2,5,8,...

$12^{*}2 + 7 \rightarrow 12^{*}3 + 5,$
$12^{*}5 + 7 \rightarrow 12^{*}7 + 5,$
$12^{*}8 + 7 \rightarrow 12^{*}10 + 5,$

Note that all natural integers that stand at position number 2 in the considered reverse horizontal sequences can be represented as $12*r+5$, where $r = v*4^{\wedge}s+t$, at $v = 0,1,2,.....$ and $t < 4^{\wedge}s$, where $s = 1$.

Hence, in all sequences that begin in elements of type 7, $12*p+7$, at $p = 2,5,8,..$ in position 2 have an element of type 5, which has an image in the reverse horizontal transformation, and therefore all the reverse horizontal sequences have a length greater than 2

Therefore, the proportion of sequences that begin in numbers 12*p+7 and have a length greater than 2 is 2/3 of the set of sequences that begin in natural integer of type 12*p+7 and whose length is greater than 1

Consider reverse horizontal sequences that begin in natural integers of type 11, where p = 0,3,6,...

12*0 + 11 → 12*0 + 7,
12*3 + 11 → 12*2 + 7,
12*6 + 11 → 12*4 + 7,

Note that all natural integers that stand at position number 2 in the considered reverse horizontal sequences can be represented as 12*r+7, where r = v*2^s+t, at v = 0,1,2,..... and t < 2^s, where s = 1.

Hence all sequences that begin in elements of type 11, 12*p+11, at p = 0,3,6,.. in position 2 have an element of type 7 element, which has image in the reverse horizontal transformation, and hence all those reverse horizontal sequences have a length greater than 2

Consider reverse horizontal sequences that begin in natural integers of type 11, where p = 1,4,7,...

12*1 + 11 → 12*1 + 3,
12*4 + 11 → 12*3 + 3,
12*7 + 11 → 12*5 + 3,

Note that all natural integers that stand at position number 2 in the considered reverse horizontal sequences can be represented as 12*r+3, where r = v*2^s+t, at v = 0,1,2,..... and t < 2^s, where s = 1.

Hence all sequences that begin in elements of type 11, 12*p+11, at p = 1,4,7,.. in position 2 have an element of type 3 element, which has no image in the reverse horizontal transformation, and hence all those reverse horizontal sequences are of length 2

Consider reverse horizontal sequences that begin in natural integers of type 11, where p = 2,5,8,...

$12^*2 + 11 \rightarrow 12^*1 + 11,$
$12^*5 + 11 \rightarrow 12^*3 + 11,$
$12^*8 + 11 \rightarrow 12^*5 + 11,$

Note that all natural integers that stand at position number 2 in the considered reverse horizontal sequences can be represented as 12^*r+11, where $r = v^*2^\wedge s+t$, at $v = 0,1,2,.....$ and $t < 2^\wedge s$, where $s = 1$.

Hence all sequences that begin in elements of type 11, 12*p+11, at p = 2,5,8,... in position 2 have an element of type 11 element, which has image in the reverse horizontal transformation, and hence all those reverse horizontal sequences have a length greater than 2

Therefore, the proportion of sequences that begin in numbers 12*p+11 and have a length greater than 2 is 2/3 of the set of sequences that begin in natural integer of type 12*p+11 and whose length is greater than 1

Consider reverse horizontal sequences that begin in natural integers of type 1, where p = 1,4,7,...

$12^*1 + 1 \rightarrow 12^*1 + 5,$

$12^*4 + 1 \rightarrow 12^*5 + 5,$

$12^*7 + 1 \rightarrow 12^*9 + 5,$

Note that all natural integers that stand at position number 2 in the considered reverse horizontal sequences can be represented as $12*r+5$, where $r = v*4^\wedge s + t$, at $v = 0,1,2,.....$ and $t < 4^\wedge s$, where $s = 1$.

Hence, in all sequences that begin in elements of type 1, $12*p+1$, at $p = 1,4,7,..$ in position 2 have an element of type 5, which has an image in the reverse horizontal transformation, and therefore all the reverse horizontal sequences have a length greater than 2

Consider reverse horizontal sequences that begin in natural integers of type 1, where $p = 2,5,8,...$

$12^*2 + 1 \rightarrow 12^*2 + 9,$

$12^*5 + 1 \rightarrow 12^*6 + 9,$

$12^*8 + 1 \rightarrow 12^*10 + 9,$

Note that all natural integers that stand at position number 2 in the considered reverse horizontal sequences can be represented as $12*r+9$, where $r = v*4^\wedge s + t$, at $v = 0,1,2,.....$ and $t < 4^\wedge s$, where $s = 1$.

Hence all sequences that begin in elements of type 1, $12*p+1$, at $p = 2,5,8,..$ in position 2 have an element type 9, which has no image in the reverse horizontal transformation, and hence all those reverse horizontal sequences are of length 2

Consider reverse horizontal sequences that begin in natural

integers of type 1, where p = 3,6,9,...

$12^\star 3 + 1 \rightarrow 12^\star 4 + 1,$
$12^\star 6 + 1 \rightarrow 12^\star 8 + 1,$
$12^\star 9 + 1 \rightarrow 12^\star 12 + 1,$

Note that all natural integers that stand at position number 2 in the considered reverse horizontal sequences can be represented as $12*r+1$, where $r = v*4^\wedge s+t$, at $v = 0,1,2,.....$ and $t < 4^\wedge s$, where $s = 1$.

Hence, in all sequences that begin in elements of type 1, $12*p+1$, at $p = 3,6,9,..$ in position 2 have an element of type 1, which has an image in the reverse horizontal transformation, and therefore all the reverse horizontal sequences have a length greater than 2

Therefore, the proportion of sequences that begin in numbers $12*p+1$ and have a length greater than 2 is 2/3 of the set of sequences that begin in natural integer of type $12*p+1$ and whose length is greater than 1

Therefore, the proportion of reverse horizontal sequences whose length is greater than 2 is $(2/3)^\wedge 2$ of the set of all reverse horizontal sequences

Note that all natural integers that stand at position number 2 in the considered horizontal sequences can be represented as $12*r+q$, where $q = 3,5,11$ and $r = v*2^\wedge s+t$, at $v = 0,1,2,.....$ and $t < 2^\wedge s$, and can be represented as $12*r+q$, where $q = 7,9,1$ and $r = v*2*2^\wedge s+t$, at $v = 0,1,2,....$ and $t < 2*2^\wedge s$, where $s = 1$.

From the above reasoning and section 12.2, it follows that

for any p > = 0 if q = 3,5,7,9,11, and p > 0 if q =1, and any
m > 0 in the set of direct horizontal sequences

12*p+q, 12*p+(q+2), 12*p+(q+4), 12*p+(q+6),
12*p+(q+8), 12*p+(q+10),
12*(p+1)+q, 12*(p+1)+(q+2), 12*(p+1)+(q+4),
12*(p+1)+(q+6),
12*(p+1)+(q+8),12*(p+1)+(q+10),....,12*(p+3*3^m-1)+q,
12*(p+3*3^m-1)+(q+2), 12*(p+3*3^m-1)+(q+4),
12*(p+3*3^m-1)+(q+6), 12*(p+3*3^m-1)+(q+8),
12*(p+3*3^m-1)+(q+10)

exactly (6*3*3^m)*(2/3)^2 are longer than 2,

namely 24 of 54, 72 of 162 and etc

Hence (see section 12.2) the proportion of reverse
horizontal sequences whose length is greater than 2 in any
set (the size of which is a multiple 6*3*3^2) reverse
horizontal sequences, which are arranged in a row, equal to
(2/3)^2

13.2.3 Rule (2/3)^n for sequences whose lengths are greater than n in a set of reverse horizontal sequences.

In the following sections from 13.2.3.1 to 13.2.3.5 it will be
proved that the proportion of reverse horizontal sequences
whose length is greater than n is (2/3)^n of the set of all
reverse horizontal sequences

13.2.3.1 Reverse horizontal sequences in which the members n, n>1 are natural integers of type 3

The reverse horizontal sequences in which the natural

integers of type 3 are at position n end in these numbers because the natural integers of type 3 have no images in the reverse horizontal transformation (see section 11)

Hence all reverse horizontal sequences that end in numbers $12*p + 3$, p=0,1,2,.. have length n

13.2.3.2 Reverse horizontal sequences in which the members n, n>1 are natural integers of type 5

Consider reverse horizontal sequences that begin in terms of $12*p+q$ and in which the terms at position n, n>1 are natural integers of type 5

$$12^*p+q \rightarrow .. \rightarrow 12^*r+5$$

let $r = v*2\wedge s+t$, where $v = 0,1,2,....$ and $t < 2\wedge s$

The following pairs of natural integers v and t are possible,

$v = 3*y + 0, t = 3*z + 0$
$v = 3*y + 0, t = 3*z + 1$
$v = 3*y + 0, t = 3*z + 2$
$v = 3*y + 1, t = 3*z + 0$
$v = 3*y + 1, t = 3*z + 1$
$v = 3*y + 1, t = 3*z + 2$
$v = 3*y + 2, t = 3*z + 0$
$v = 3*y + 2, t = 3*z + 1$
$v = 3*y + 2, t = 3*z + 2$

let $v = 3*y + 0, t = 3*z + 0$ then

$$12^*(v^*2\wedge s+t)+5 = 12^*3^*y^*2\wedge s+12^*3^*z+5 \rightarrow$$

$$(12*(3*y*2^\wedge(s+1)+3*z*2)+2*5 - 1)/3 =$$
$$12*(y*2^\wedge(s+1)+z*2)+3$$

that is, the images of the numbers $12*(v*2^\wedge s+t)+5$ in the reverse horizontal transformation will be the numbers, $12*(y*2^\wedge(s+1)+z*2)+3$.

Let us note that

$$3*z < 2^\wedge s,$$
$$2*z < 2^\wedge s,$$
$$2*z < 2^\wedge(s+1)$$

and hence the numbers that stand at position n+1 in the considered reverse horizontal sequences can be represented as $12*r+3$ and $r = v*2^\wedge(s+1)+t$, at $v = 0,1,2,.....$ and $t < 2^\wedge(s+1)$

let $v = 3*y + 0$, $t = 3*z + 1$ then

$$12*(v*2^\wedge s+t)+5 = 12*3*y*2^\wedge s+12*3*z+12+5 \rightarrow$$
$$(12*(3*y*2^\wedge(s+1)+3*z*2)+2*17 - 1)/3 =$$
$$12*(y*2^\wedge(s+1)+z*2)+11$$

that is, the images of the numbers $12*(v*2^\wedge s+t)+5$ in the reverse horizontal transformation will be the numbers, $12*(y*2^\wedge(s+1)+z*2)+11$.

Let us note that

$$3*z + 1 < 2^\wedge s,$$
$$3*z < 2^\wedge s,$$
$$2*z < 2^\wedge s,$$
$$2*z < 2^\wedge(s+1)$$

and hence the numbers that stand at position n+1 in the considered reverse horizontal sequences can be represented as 12*r+11 and r = v*2^(s+1)+t, at v = 0,1,2,.... and t < 2^(s+1)

let v = 3*y + 0, t = 3*z + 2
then

$12^*(v^*2^\wedge s+t)+5 = 12^*3^*y^*2^\wedge s+12^*3^*z+12^*2+5 \rightarrow$
$(12^*(3^*y^*2^\wedge(s+1)+3^*z^*2)+2^*29 - 1)/3 =$
$12^*(y^*2^\wedge(s+1)+(z^*2+1))+7$

that is, the images of the numbers 12*(v*2^s+t)+5 in the reverse horizontal transformation will be the numbers, 12*(y*2^(s+1)+(z*2+1))+7.

Let us note that

$3^*z + 2 < 2^\wedge s,$
$3^*z + 1 < 2^\wedge s,$
$2^*z + 1 < 2^\wedge s,$
$2^*z + 1 < 2^\wedge(s+1)$

and hence the numbers that stand at position n+1 in the considered reverse horizontal sequences can be represented as 12*r+7 and r = v*2^(s+1)+t, at v = 0,1,2,.... and t < 2^(s+1)

let v = 3*y + 1, t = 3*z + 0 then

$12^*(v^*2^\wedge s+t)+5 = 12^*(3^*y+1)^*2^\wedge s+12^*3^*z+5 \rightarrow$
$(12^*(3^*y^*2^\wedge(s+1)+2^\wedge(s+1)+3^*z^*2)+2^*5 - 1)/3 =$

108

$$= 12*(y*2^{(s+1)}+z*2))+4*2^{(s+1)}+3$$

natural integers 2^s can be represented as

$$2^s = 12*m1 + 4, s = 2,4,6,\ldots$$

or

$$2^s = 12*m2 + 8, s = 3,5,7,\ldots$$

accordingly natural integers $4*2^{(s+1)}$ can be represented as

$$4*2^{(s+1)} = 12*m3 + 8, s = 2,4,6,\ldots$$

or

$$4*2^{(s+1)} = 12*m4 + 4, s = 3,5,7,\ldots$$

accordingly natural integers $4*2^{(s+1)}+3$ can be represented as

$$4*2^{(s+1)}+3 = 12*m5 + 11, s = 2,4,6,\ldots$$

or

$$4*2^{(s+1)}+3 = 12*m6 + 7, s = 3,5,7,\ldots$$

accordingly

$$12*(y*2^{(s+1)}+z*2))+4*2^{(s+1)}+3 =$$
$$12*(y*2^{(s+1)}+z*2+m5)+11, s = 2,4,6,\ldots$$

or

$$12*(y*2^{(s+1)}+z*2))+4*2^{(s+1)}+3 =$$

$12*(y*2^{(s+1)}+z*2+m6)+7$, s = 3,5,7,…

Let us note that

$m5 = (4*2^{(s+1)}-8)/12 = (2^{(s+1)}-2)/3$, s = 2,4,6,…

$3*z < 2^s$,
$3*z < 2^{(s+1)}/2$,
$2*z < 2^{(s+1)}/2$,
$(2^{(s+1)}-2)/3 < 2^{(s+1)}/3$,
$(2^{(s+1)}-2)/3 + 2*z < 2^{(s+1)}/3 + 2^{(s+1)}/2$
$(2^{(s+1)}-2)/3 + z*2 < 2^{(s+1)}$

and

$m6 = (4*2^{(s+1)}-4)/12 = (2^{(s+1)}-1)/3$, s = 3,5,7,…

$3*z < 2^s$,
$3*z < 2^{(s+1)}/2$,
$2*z < 2^{(s+1)}/2$,
$(2^{(s+1)}-1)/3 < 2^{(s+1)}/3$,
$(2^{(s+1)}-1)/3 + 2*z < 2^{(s+1)}/3 + 2^{(s+1)}/2$
$(2^{(s+1)}-1)/3 + z*2 < 2^{(s+1)}$

and hence the numbers that stand at position n+1 in the considered reverse horizontal sequences can be represented as $12*r+11$ and $r = v*2^{(s+1)}+t$, at v = 0,1,2,…., s = 2,4,6,… and $t < 2^{(s+1)}$ or $12*r+7$ and $r = v*2^{(s+1)}+t$, at v = 0,1,2,…., s = 3,5,7,… and $t < 2^{(s+1)}$

let v = 3*y + 1, t = 3*z + 1 then

$12*(v*2^s+t)+5 = 12*(3*y+1)*2^s + 12*3*z+12+5 \rightarrow$
$(12*(3*y*2^{(s+1)}+2^{(s+1)}+3*z*2)+2*17 - 1)/3 =$

$$12*(y*2^{\wedge}(s+1)+z*2)+4*2^{\wedge}(s+1)+11$$

natural integers $2^{\wedge}s$ can be represented as

$$2^{\wedge}s = 12*m1 + 4, s = 2,4,6,\ldots$$

or

$$2^{\wedge}s = 12*m2 + 8, s = 3,5,7,\ldots$$

accordingly natural integers $4*2^{\wedge}(s+1)$ can be represented as

$$4*2^{\wedge}(s+1) = 12*m3 + 8, s = 2,4,6,\ldots$$

or

$$4*2^{\wedge}(s+1) = 12*m4 + 4, s = 3,5,7,\ldots$$

accordingly natural integers $4*2^{\wedge}(s+1)+11$ can be represented as

$$4*2^{\wedge}(s+1)+11 = 12*m5 + 7, s = 2,4,6,\ldots$$

or

$$4*2^{\wedge}(s+1)+11 = 12*m6 + 3, s = 3,5,7,\ldots$$

accordingly

$$12*(y*2^{\wedge}(s+1)+z*2))+4*2^{\wedge}(s+1)+11 =$$
$$12*(y*2^{\wedge}(s+1)+z*2+m5)+7, s = 2,4,6,\ldots$$

or

$$12*(y*2^{\wedge}(s+1)+z*2))+4*2^{\wedge}(s+1)+11 =$$

12*(y*2^(s+1)+z*2+m6)+3, s = 3,5,7,...

Let us note that

m5 = (4*2^(s+1)+4)/12 = (2^(s+1)+1)/3, s = 2,4,6,...

3*z+1 < 2^s,
3*z < 2^(s+1)/2 - 1,
2*z < 2^(s+1)/2 - 1,
(2^(s+1)+1)/3 < 2^(s+1)/3+ 1,
(2^(s+1)+1)/3 + 2*z < 2^(s+1)/3+ 1 + 2^(s+1)/2 - 1
(2^(s+1)+1)/3+z*2 < 2^(s+1)

and

m6 = (4*2^(s+1)+8)/12 = (2^(s+1)+2)/3, s = 3,5,7,...

3*z+1 < 2^s,
3*z < 2^(s+1)/2 - 1,
2*z < 2^(s+1)/2 - 1,
(2^(s+1)+2)/3 < 2^(s+1)/3 + 1,
(2^(s+1)+2)/3 + 2*z < 2^(s+1)/3 + 1+ 2^(s+1)/2 - 1
(2^(s+1)+2)/3+z*2 < 2^(s+1)

and hence the numbers that stand at position n+1 in the considered reverse horizontal sequences can be represented as 12*r+7 and r = v*2^(s+1)+t, at v = 0,1,2,...., s = 2,4,6,... and t < 2^(s+1) or 12*r+3 and r = v*2^(s+1)+t, at v = 0,1,2,...., s = 3,5,7,... and t < 2^(s+1)

let v = 3*y + 1, t = 3*z + 2 then

12*(v*2^s+t)+5 = 12*(3*y+1)*2^s + 12*3*z+2*12+5 → =
(12*(3*y*2^(s+1)+3*z*2)+12*2^(s+1)+2*29 − 1)/3 =

$$12*(y*2^{\wedge}(s+1)+(z*2+1))+4*2^{\wedge}(s+1)+7$$

natural integers $2^{\wedge}s$ can be represented as

$$2^{\wedge}s = 12*m1 + 4,\ s = 2,4,6,\ldots$$

or

$$2^{\wedge}s = 12*m2 + 8,\ s = 3,5,7,\ldots$$

accordingly natural integers $4*2^{\wedge}(s+1)$ can be represented as

$$4*2^{\wedge}(s+1) = 12*m3 + 8,\ s = 2,4,6,\ldots$$

or

$$4*2^{\wedge}(s+1) = 12*m4 + 4,\ s = 3,5,7,\ldots$$

accordingly natural integers $4*2^{\wedge}(s+1)+7$ can be represented as

$$4*2^{\wedge}(s+1)+7 = 12*m5 + 3,\ s = 2,4,6,\ldots$$

or

$$4*2^{\wedge}(s+1)+7 = 12*m6 + 11,\ s = 3,5,7,\ldots$$

accordingly

$$12*(y*2^{\wedge}(s+1)+(z*2+1))+4*2^{\wedge}(s+1)+7 =$$
$$12*(y*2^{\wedge}(s+1)+(z*2+1)+m5)+3,\ s = 2,4,6,\ldots$$

or

$$12*(y*2^{\wedge}(s+1)+(z*2+1))+4*2^{\wedge}(s+1)+3 =$$

113

$12*(y*2^{(s+1)}+(z*2+1)+m6)+11$, s = 3,5,7,...

Let us note that

$m5 = (4*2^{(s+1)}+4)/12 = (2^{(s+1)}+1)/3$, s = 2,4,6,...

$3*z+2 < 2^s$,
$3*z < 2^{(s+1)}/2 - 2$,
$2*z < 2^{(s+1)}/3 - 4/3$,
$2*z +1 < 2^{(s+1)}/3 - 1/3$,
$(2^{(s+1)}+1)/3 = 2^{(s+1)}/3 + 1/3$,
$(2^{(s+1)}+1)/3 + (2*z +1) < 2^{(s+1)}/3 + 1/3 + 2^{(s+1)}/3 - 1/3$
$(2^{(s+1)}+1)/3 + (z*2+1) < 2^{(s+1)}$

and

$m6 = (4*2^{(s+1)}-4)/12 = (2^{(s+1)}-1)/3$, s = 3,5,7,...

$3*z+2 < 2^s$,
$3*z < 2^{(s+1)}/2 - 2$,
$2*z < 2^{(s+1)}/3 - 4/3$,
$2*z +1 < 2^{(s+1)}/3 - 1/3$,
$(2^{(s+1)}-1)/3 = 2^{(s+1)}/3 - 1/3$,
$(2^{(s+1)}-1)/3 + (2*z+1) < 2^{(s+1)}/3 + 2^{(s+1)}/3 - 2/3$
$(2^{(s+1)}-1)/3 + (z*2+1) < 2^{(s+1)}$

and hence the numbers that stand at position n+1 in the considered reverse horizontal sequences can be represented as $12*r+3$ and $r = v*2^{(s+1)}+t$, at v = 0,1,2,...., s = 2,4,6,... and $t < 2^{(s+1)}$ or $12*r+11$ and $r = v*2^{(s+1)}+t$, at v = 0,1,2,...., s = 3,5,7,... and $t < 2^{(s+1)}$

let $v = 3*y + 2$, $t = 3*z + 0$ then

$12^{*}(v^{*}2^{\wedge}s+t)+5 = 12^{*}(3^{*}y+2)^{*}2^{\wedge}s+12^{*}3^{*}z+5 \rightarrow$
$(12^{*}(3^{*}y^{*}2^{\wedge}(s+1)+2^{*}2^{\wedge}(s+1)+3^{*}z^{*}2)+2^{*}5 - 1)/3 =$

$= 12^{*}(y^{*}2^{\wedge}(s+1))+z^{*}2)+4^{*}2^{*}2^{\wedge}(s+1)+3$

natural integers $2^{\wedge}s$ can be represented as

$2^{\wedge}s = 12^{*}m1 + 4$, $s = 2,4,6,...$

or

$2^{\wedge}s = 12^{*}m2 + 8$, $s = 3,5,7,...$

accordingly natural integers $4^{*}2^{*}2^{\wedge}(s+1)$ can be represented as

$4^{*}2^{*}2^{\wedge}(s+1) = 12^{*}m3 + 4$, $s = 2,4,6,...$

or

$4^{*}2^{*}2^{\wedge}(s+1) = 12^{*}m4 + 8$, $s = 3,5,7,...$

accordingly natural integers $4^{*}2^{*}2^{\wedge}(s+1)+3$ can be represented as

$4^{*}2^{*}2^{\wedge}(s+1)+3 = 12^{*}m5 + 7$, $s = 2,4,6,...$

or

$4^{*}2^{*}2^{\wedge}(s+1)+3 = 12^{*}m6 + 11$, $s = 3,5,7,...$

accordingly

$12^{*}(y^{*}2^{\wedge}(s+1)+z^{*}2))+4^{*}2^{*}2^{\wedge}(s+1)+3 =$
$12^{*}(y^{*}2^{\wedge}(s+1)+z^{*}2+m5)+7$, $s = 2,4,6,...$

or

$$12*(y*2^{(s+1)}+z*2))+4*2*2^{(s+1)}+3 =$$
$$12*(y*2^{(s+1)}+z*2+m6)+11, \ s = 3,5,7,\ldots$$

Let us note that

$$m5 = (4*2*2^{(s+1)}-4)/12 = (2*2^{(s+1)}-1)/3, \ s = 2,4,6,\ldots$$

$$3*z < 2^s,$$
$$3*z < 2^{(s+1)}/2,$$
$$2*z < 2^{(s+1)}/3,$$
$$(2*2^{(s+1)})/3 = (2*2^{(s+1)})/3,$$
$$(2*2^{(s+1)}-1)/3 < (2*2^{(s+1)})/3,$$
$$(2*2^{(s+1)}-1)/3+z*2 < (2*2^{(s+1)})/3 + 2^{(s+1)}/3,$$
$$(2*2^{(s+1)}-1)/3+z*2 < 2^{(s+1)}$$

and

$$m6 = (4*2*2^{(s+1)}-8)/12 = (2*2^{(s+1)}-2)/3, \ s = 3,5,7,\ldots$$

$$3*z < 2^s,$$
$$3*z < 2^{(s+1)}/2,$$
$$2*z < 2^{(s+1)}/3,$$
$$(2*2^{(s+1)})/3 = (2*2^{(s+1)})/3,$$
$$(2*2^{(s+1)}-2)/3 < (2*2^{(s+1)})/3,$$
$$(2*2^{(s+1)}-2)/3+z*2 < (2*2^{(s+1)})/3 + 2^{(s+1)}/3,$$
$$(2*2^{(s+1)}-2)/3+z*2 < 2^{(s+1)}$$

and hence the numbers that stand at position n+1 in the considered reverse horizontal sequences can be represented as 12*r+7 and r = v*2^{(s+1)}+t, at v = 0,1,2,...., s = 2,4,6,... and t < 2^{(s+1)} or 12*r+11 and r = v*2^{(s+1)}+t, at v = 0,1,2,...., s = 3,5,7,... and t < 2^{(s+1)}

let $v = 3*y + 2$, $t = 3*z + 1$ then

$12*(v*2^s+t)+5 = 12*3*y*2^s+12*2*2^s+ 12*3*z+12+5 \rightarrow$
$(12*(3*y*2^{(s+1)}+3*z*2)+12*2*2^{(s+1)}+2*17 - 1)/3 =$
$12*(y*2^{(s+1)}+z*2)+4*2*2^{(s+1)} +11$

natural integers 2^s can be represented as

$2^s = 12*m1 + 4$, $s = 2,4,6,\ldots$

or

$2^s = 12*m2 + 8$, $s = 3,5,7,\ldots$

accordingly natural integers $4*2*2^{(s+1)}$ can be represented as

$4*2*2^{(s+1)} = 12*m3 + 4$, $s = 2,4,6,\ldots$

or

$4*2*2^{(s+1)} = 12*m4 + 8$, $s = 3,5,7,\ldots$

accordingly natural integers $4*2^{(s+1)}+11$ can be represented as

$4*2*2^s +11 = 12*m5 + 3$, $s = 2,4,6,\ldots$

or

$4*2*2^s +11 = 12*m6 + 7$, $s = 3,5,7,\ldots$

accordingly

$12*(y*2^{(s+1)}+z*2))+4*2*2^s +11 =$

12*(y*2^(s+1)+z*2+m5)+3, s = 2,4,6,...

or

12*(y*2^(s+1)+z*2))+4*2*2^s +11 =
12*(y*2^(s+1)+z*2+m6)+7, s = 3,5,7,...

Let us note that

m5 = (4*2*2^(s+1)+8)/12 = (2*2^(s+1)+2)/3, s = 2,4,6,...

3*z+1 < 2^s,
3*z +1< 2^(s+1)/2,
3*z < 2^(s+1)/2 - 1,
2*z < 2^(s+1)/3 − 2/3,
(2*2^(s+1))/3 = (2*2^(s+1))/3,
(2*2^(s+1)+2)/3 = (2*2^(s+1))/3+2/3,
(2*2^(s+1)+2)/3+z*2 < (2*2^(s+1))/3+2/3 + 2^(s+1)/3 −
2/3,
(2*2^(s+1)+2)/3+z*2 < 2^(s+1)

and

m6 = (4*2*2^(s+1)+4)/12 = (2*2^(s+1)+1)/3, s = 3,5,7,...

3*z+1 < 2^s,
3*z +1< 2^(s+1)/2,
3*z < 2^(s+1)/2 - 1,
2*z < 2^(s+1)/3 − 2/3,
(2*2^(s+1))/3 = (2*2^(s+1))/3,
(2*2^(s+1)+1)/3 < (2*2^(s+1))/3+2/3,
(2*2^(s+1)+1)/3+z*2 < (2*2^(s+1))/3 + 2/3+ 2^(s+1)/3 −
2/3,
(2*2^(s+1)+1)/3+z*2 < 2^(s+1)

and hence the numbers that stand at position n+1 in the considered reverse horizontal sequences can be represented as $12*r+3$ and $r = v*2^{(s+1)}+t$, at $v = 0,1,2,....,$ $s = 2,4,6,...$ and $t < 2^{(s+1)}$ or $12*r+7$ and $r = v*2^{(s+1)}+t$, at $v = 0,1,2,....,$ $s = 3,5,7,...$ and $t < 2^{(s+1)}$

let $v = 3*y + 2$, $t = 3*z + 2$ then

$$12^*(v^*2\hat{\ }s+t)+5 = 12^*(3^*y+2)^*2\hat{\ }s+ 12^*3^*z+2^*12+5 \rightarrow =$$
$$(12^*(3^*y^*2\hat{\ }(s+1)+3^*z^*2)+12^*2^*2\hat{\ }(s+1)+2^*29 - 1)/3 =$$
$$12^*(y^*2\hat{\ }(s+1)+(z^*2+1))+4^*2^*2\hat{\ }(s+1)+7$$

natural integers $2\hat{\ }s$ can be represented as

$2\hat{\ }s = 12*m1 + 4$, $s = 2,4,6,...$

or

$2\hat{\ }s = 12*m2 + 8$, $s = 3,5,7,...$

accordingly natural integers $4*2*2\hat{\ }(s+1)$ can be represented as

$4*2*2\hat{\ }(s+1) = 12*m3 + 4$, $s = 2,4,6,...$

or

$4*2*2\hat{\ }(s+1) = 12*m4 + 8$, $s = 3,5,7,...$

accordingly natural integers $4*2\hat{\ }(s+1)+7$ can be represented as

$4*2*2\hat{\ }s + 7 = 12*m5 + 11$, $s = 2,4,6,...$

or

$4*2*2^s + 7 = 12*m6 + 3$, s = 3,5,7,...

accordingly

$12*(y*2^{(s+1)}+z*2+1)+4*2*2^s +7 =$
$12*(y*2^{(s+1)}+(z*2+1)+m5)+11$, s = 2,4,6,...

or

$12*(y*2^{(s+1)}+z*2+1)+4*2*2^s +7 =$
$12*(y*2^{(s+1)}+(z*2+1)+m6)+3$, s = 3,5,7,...

Let us note that

$m5 = (4*2*2^{(s+1)}-4)/12 = (2*2^{(s+1)}-1)/3$, s = 2,4,6,...

$3*z+2 < 2^s$,
$3*z < 2^{(s+1)}/2 - 2$,
$2*z < 2^{(s+1)}/3 - 4/3$,
$2*z +1 < 2^{(s+1)}/3 - 1/3$,
$(2*2^{(s+1)} - 1)/3 = (2*2^{(s+1)})/3 - 1/3$,
$(2*2^{(s+1)}-1)/3 < (2*2^{(s+1)})/3$,
$(2*2^{(s+1)}-1)/3+(z*2+1) < (2*2^{(s+1)})/3 + 2^{(s+1)}/3 -2/3$,
$(2*2^{(s+1)}-1)/3+(z*2+1) < 2^{(s+1)}$

and

$m6 = (4*2*2^{(s+1)}+4)/12 = (2*2^{(s+1)}+1)/3$, s = 3,5,7,...

$3*z+2 < 2^s$,
$3*z < 2^{(s+1)}/2 - 2$,
$2*z < 2^{(s+1)}/3 - 4/3$,
$2*z +1 < 2^{(s+1)}/3 - 1/3$,

$(2*2^\wedge(s+1)+1)/3 = 2*2^\wedge(s+1)/3+1/3,$

$(2*2^\wedge(s+1)+1)/3+(z*2+1) < (2*2^\wedge(s+1))/3+ 1/3 + 2^\wedge(s+1)/3 - 1/3,$

$(2*2^\wedge(s+1)+1)/3+(z*2+1) < 2^\wedge(s+1)$

and hence the numbers that stand at position n+1 in the considered reverse horizontal sequences can be represented as 12*r+11 and r = v*2^(s+1)+t, at v = 0,1,2,...., s = 2,4,6,... and t < 2^(s+1) or 12*r+3 and r = v*2^(s+1)+t, at v = 0,1,2,...., s = 3,5,7,... and t < 2^(s+1)

13.2.3.3 Reverse horizontal sequences in which the members n, n>1 are natural integers of type 7

Consider reverse horizontal sequences that begin in terms of 12*p+q and in which the terms at position n, n>1 are natural integers of type 7

$12^*p+q \rightarrow .. \rightarrow 12^*r+7$

let r = v*2^s+t, where v = 0,1,2,.... and t <2^s

The following pairs of natural integers v and t are possible,

v = 3*y + 0, t = 3*z + 0
v = 3*y + 0, t = 3*z + 1
v = 3*y + 0, t = 3*z + 2
v = 3*y + 1, t = 3*z + 0
v = 3*y + 1, t = 3*z + 1
v = 3*y + 1, t = 3*z + 2
v = 3*y + 2, t = 3*z + 0
v = 3*y + 2, t = 3*z + 1
v = 3*y + 2, t = 3*z + 2

let $v = 3*y + 0$, $t = 3*z + 0$ then

$12^*(v*2^{\wedge}s+t)+7 = 12^*(3^*y+0)^*2^{\wedge}s+12^*(3^*z+0)+7 \rightarrow$
$(12^*(3^*y^*4^*2^{\wedge}s+3^*z^*4)+4^*7-1)/3 =$
$12^*(y^*2^*2^{\wedge}(s+1)+z^*4)+9$

Let us note that

$3^*z < 2^{\wedge}s$,
$2^*z < 2^{\wedge}s$,
$2^*2^*z < 2^{\wedge}(s+1)$,
$4^*z < 2^{\wedge}(s+1)$
$4^*z < 2^*2^{\wedge}(s+1)$

and hence the numbers that stand at position n+1 in the considered reverse horizontal sequences can be represented as 12^*r+9 and $r = v^*2^*2^{\wedge}(s+1)+t$, at $v = 0,1,2,....$ and $t < 2^*2^{\wedge}(s+1)$

let $v = 3*y + 0$, $t = 3*z + 1$ then

$12^*(v*2^{\wedge}s+t)+7 = 12^*(3^*y+0)^*2^{\wedge}s+12^*(3^*z+1)+7 \rightarrow$
$(12^*(3^*y^*2^*2^{\wedge}(s+1)+3^*z^*4)+4^*12+4^*7-1)/3 =$
$12^*(y^*2^*2^{\wedge}(s+1)+(z^*4+2))+1$

Let us note that

$3^*z+1 < 2^{\wedge}s$,
$2^*z+1 < 2^{\wedge}s$,
$2^*2^*z+2 < 2^{\wedge}(s+1)$,
$4^*z+2 < 2^{\wedge}(s+1)$
$4^*z+2 < 2^*2^{\wedge}(s+1)$

and hence the numbers that stand at position n+1 in the

considered reverse horizontal sequences can be represented as 12*r+1 and r = v*2*2^(s+1)+t, at v = 0,1,2,…. and t < 2*2^(s+1)

let v = 3*y + 0, t = 3*z + 2 then

$12^{\star}(v^{\star}2^{\wedge}s+t)+7 = 12^{\star}(3^{\star}y+0)^{\star}2^{\wedge}s+12^{\star}(3^{\star}z+2)+7 \rightarrow$
$(12^{\star}(3^{\star}y^{\star}2^{\star}2^{\wedge}(s+1)+3^{\star}z^{\star}4)+4^{\star}12^{\star}2+4^{\star}7 - 1)/3 =$
$= 12^{\star}(y^{\star}2^{\star}2^{\wedge}(s+1)+(z^{\star}4+3))+5$

Let us note that

$3^{\star}z+2 < 2^{\wedge}s,$
$2^{\star}z+2 < 2^{\wedge}s,$
$2^{\star}2^{\star}z+4 < 2^{\wedge}(s+1),$
$4^{\star}z+3 < 2^{\wedge}(s+1)$
$4^{\star}z+3 < 2^{\star}2^{\wedge}(s+1)$

and hence the numbers that stand at position n+1 in the considered reverse horizontal sequences can be represented as 12*r+5 and r = v*2*2^(s+1)+t, at v = 0,1,2,…. and t < 2*2^(s+1)

let v = 3*y + 1, t = 3*z + 0 then

$12^{\star}(v^{\star}2^{\wedge}s+t)+7 = 12^{\star}(3^{\star}y+1)^{\star}2^{\wedge}s+12^{\star}3^{\star}z+7 \rightarrow$
$(12^{\star}(3^{\star}y^{\star}2^{\star}2^{\wedge}(s+1)+2^{\star}2^{\wedge}(s+1)+3^{\star}z^{\star}4)+4^{\star}7 - 1)/3 =$
$= 12^{\star}(y^{\star}2^{\star}2^{\wedge}(s+1)+z^{\star}4))+4^{\star}2^{\star}2^{\wedge}(s+1)+9$

natural integers 2^s can be represented

$2^{\wedge}s = 12^{\star}m1 + 4, s = 2,4,6,\ldots$

or

$$2^s = 12*m2 + 8, s = 3,5,7,...$$

accordingly natural integers $4*2*2^{(s+1)}$ can be represented

$$4*2*2^{(s+1)} = 12*m3 + 4, s = 2,4,6,...$$

or

$$4*2*2^{(s+1)} = 12*m4 + 8, s = 3,5,7,...$$

accordingly natural integers $4*2*2^{(s+1)}+9$ can be represented

$$4*2*2^{(s+1)}+9 = 12*m5 + 1, s = 2,4,6,...$$

or

$$4*2*2^{(s+1)}+9 = 12*m6 + 5, s = 3,5,7,...$$

accordingly

$$12*(y*2*2^{(s+1)}+z*4))+4*2*2^{(s+1)}+9 =$$
$$12*(y*2*2^{(s+1)}+z*4+m5)+1, s = 2,4,6,...$$

or

$$12*(y*2*2^{(s+1)}+z*4))+4*2*2^{(s+1)}+9 =$$
$$12*(y*2*2^{(s+1)}+z*4+m6)+5, s = 3,5,7,...$$

Let us note that

$$m5 = (4*2*2^{(s+1)}+8)/12 = (2*2^{(s+1)}+2)/3, s = 2,4,6,...$$

$3*z < 2^s,$
$3*z < 2^{(s+1)}/2,$
$3*z +1 =< 2^{(s+1)}/2,$
$3*z =< 2^{(s+1)}/2 -1,$
$4*z =< 2*2^{(s+1)}/3 - 4/3,$
$(2*2^{(s+1)}+2)/3 = 2*2^{(s+1)}/3 + 2/3,$
$(2*2^{(s+1)}+2)/3 + 4*z =< 2*2^{(s+1)}/3 + 2/3+ 2*2^{(s+1)}/3$
$- 4/3$
$(2*2^{(s+1)}+2)/3 + 4*z =< 4*2^{(s+1)}/3 -2/3$
$(2*2^{(s+1)}+2)/3+z*4 < 4*2^{(s+1)}/3$
$(2*2^{(s+1)}+2)/3+z*4 < 2*2^{(s+1)}$

and

$m6 = (4*2*2^{(s+1)}+4)/12 = (2*2^{(s+1)}+1)/3, s = 3,5,7,...$

$3*z < 2^s,$
$3*z < 2^{(s+1)}/2,$
$3*z +1 =< 2^{(s+1)}/2,$
$4*z =< 2*2^{(s+1)}/3 - 4/3,$
$(2*2^{(s+1)}+1)/3 = 2*2^{(s+1)}/3 + 1/3,$
$(2*2^{(s+1)}+1)/3 + 4*z =< 2*2^{(s+1)}/3+ 2*2^{(s+1)}/3 + 1/3$
$-4/3$
$(2*2^{(s+1)}+1)/3 + 4*z =< 4*2^{(s+1)}/3 - 1$
$(2*2^{(s+1)}+1)/3+z*4 < 2*2^{(s+1)}$

and hence the numbers that stand at position n+1 in the considered reverse horizontal sequences can be represented as 12*r+5 and r = $v*2*2^{(s+1)}+t$, at v = 0,1,2,...., s = 2,4,6,... and t < $2*2^{(s+1)}$ or 12*r+1 and r = $v*2*2^{(s+1)}+t$, at v = 0,1,2,...., s = 3,5,7,... and t < $2*2^{(s+1)}$

let v = 3*y + 1, t = 3*z + 1 then

$12*(v*2^s+t)+7 = 12*(3^*y+1)*2^s+12*(3^*z+1)+7 \rightarrow$
$(12*(3^*y*2*2^{(s+1)}+2*2^{(s+1)}+3^*z*4)+4*12+4*7 - 1)/3 =$
$12*(y*2*2^{(s+1)}+(z*4+2))+4*2*2^{(s+1)}+1$

natural integers 2^s can be represented

$2^s = 12*m1 + 4, s = 2,4,6,...$

or

$2^s = 12*m2 + 8, s = 3,5,7,...$

accordingly natural integers $4*2*2^{(s+1)}$ can be represented

$4*2*2^{(s+1)} = 12*m3 + 4, s = 2,4,6,...$

or

$4*2*2^{(s+1)} = 12*m4 + 8, s = 3,5,7,...$

accordingly natural integers $4*2*2^{(s+1)}+1$ can be represented

$4*2*2^{(s+1)}+1 = 12*m5 + 5, s = 2,4,6,...$

or

$4*2*2^{(s+1)}+1 = 12*m6 + 9, s = 3,5,7,...$

accordingly

$12*(y*2*2^{(s+1)}+(z*4+2))+4*2^{(s+1)}+1 =$
$12*(y*2^{(s+1)}+z*4+m5)+5, s = 2,4,6,...$

or

12*(y*2*2^(s+1)+(z*4+2))+4*2^(s+1)+1 =
12*(y*2^(s+1)+z*4+m6)+9, s = 3,5,7,...

Let us note that

m5 = (4*2^(s+1)-4)/12 = (2*2^(s+1)-1)/3, s = 2,4,6,...

3*z +1< 2^s,
3*z + 1< 2^(s+1)/2,
3*z < 2^(s+1)/2 — 1,
3*z =< 2^(s+1)/2 — 2,
4*z =< 2*2^(s+1)/3 - 8/3,
4*z + 2 =< 2*2^(s+1)/3 -2/3,
(2*2^(s+1)-1)/3 = 2*2^(s+1)/3 - 1/3,
(2*2^(s+1)-1)/3 + 4*z +2 =< 2*2^(s+1)/3+ 2*2^(s+1)/3 - 1
(2*2^(s+1)-1)/3+z*4 +2< 2*2^(s+1)

and

m6 = (4*2^(s+1)-8)/12 = (2*2^(s+1)-2)/3, s = 3,5,7,...

3*z +1< 2^s,
3*z + 1< 2^(s+1)/2,
3*z < 2^(s+1)/2 — 1,
3*z =< 2^(s+1)/2 — 2,
4*z =< 2*2^(s+1)/3 - 8/3,
4*z + 2 =< 2*2^(s+1)/3 -2/3,
(2*2^(s+1)-2)/3 = 2^(s+1)/3 -2/3,
(2*2^(s+1)-2)/3 + 4*z+2 < 2*2^(s+1)/3+ 2*2^(s+1)/3 – 4/3
(2*2^(s+1)-2)/3+z*4+2 < 2*2^(s+1)

and hence the numbers that stand at position n+1 in the considered reverse horizontal sequences can be represented as 12*r+5 and r = v*2*2^(s+1)+t, at v = 0,1,2,...., s = 2,4,6,... and t < 2*2^(s+1) or 12*r+9 and r =

$v*2*2^{\wedge}(s+1)+t$, at $v = 0,1,2,\ldots$, $s = 3,5,7,\ldots$ and $t <$
$2*2^{\wedge}(s+1)$

let $v = 3*y + 1$, $t = 3*z + 2$ then

$12^{*}(v*2^{\wedge}s+t)+7 = 12^{*}(3^{*}y+1)^{*}2^{\wedge}s+12^{*}(3^{*}z+2)+7 \rightarrow$
$(12^{*}(3^{*}y^{*}2^{*}2^{\wedge}(s+1)+2^{*}2^{\wedge}(s+1)+3^{*}z^{*}4)+4^{*}12^{*}2+4^{*}7 - 1)/3 =$
$= 12^{*}(y^{*}2^{*}2^{\wedge}(s+1)+(z^{*}4+3))+2^{*}2^{\wedge}(s+1)+5$

natural integers $2^{\wedge}s$ can be represented

$2^{\wedge}s = 12*m1 + 4$, $s = 2,4,6,\ldots$

or

$2^{\wedge}s = 12*m2 + 8$, $s = 3,5,7,\ldots$

accordingly natural integers $2*2^{\wedge}(s+1)$ can be represented

$4*2^{\wedge}(s+1) = 12*m3 + 4$, $s = 2,4,6,\ldots$

or

$4*2^{\wedge}(s+1) = 12*m4 + 8$, $s = 3,5,7,\ldots$

accordingly natural integers $2*2^{\wedge}(s+1)+5$ can be
represented

$2*2^{\wedge}(s+1)+5 = 12*m5 + 9$, $s = 2,4,6,\ldots$

or

$2*2^{\wedge}(s+1)+5 = 12*m6 + 1$, $s = 3,5,7,\ldots$

accordingly

128

$12*(y*2*2^{(s+1)}+(z*4+3))+2*2^{(s+1)}+5 =$
$12*(y*2*2^{(s+1)}+(z*4+3)+m5)+9$, $s = 2,4,6,\ldots$

or

$12*(y*2*2^{(s+1)}+(z*4+3))+2*2^{(s+1)}+5 =$
$12*(y*2*2^{(s+1)}+(z*4+3)+m6)+1$, $s = 3,5,7,\ldots$

Let us note that

$m5 = (2*2^{(s+1)}-4)/12 = (2*2^{(s+1)}-1)/3$, $s = 2,4,6,\ldots$

$3*z +2< 2^s,$
$3*z + 2< 2^{(s+1)}/2,$
$3*z < 2^{(s+1)}/2 - 2,$
$4*z < 2*2^{(s+1)}/3 — 8/3,$
$4*z +3< 2*2^{(s+1)}/3 +1/3,$
$(2*2^{(s+1)}-1)/3 = 2*2^{(s+1)}/3 -1/3,$
$(2*2^{(s+1)}-1)/3 + 4*z+3 < 2*2^{(s+1)}/3+ 2*2^{(s+1)}/3$
$(2*2^{(s+1)}-1)/3+z*4 +3 < 2*2^{(s+1)})$

and

$m6 = (2*2^{(s+1)}+4)/12 = (2*2^{(s+1)}+1)/3$, $s = 3,5,7,\ldots$

$3*z +2< 2^s,$
$3*z + 2< 2^{(s+1)}/2,$
$3*z < 2^{(s+1)}/2 - 2,$
$4*z < 2*2^{(s+1)}/3 — 8/3,$
$4*z +3< 2*2^{(s+1)}/3 +1/3,$
$(2*2^{(s+1)}+1)/3 = 2*2^{(s+1)}/3 + 1/3,$
$(2*2^{(s+1)}+1)/3 + 4*z +3< 2^{(s+1)}/3+ 2*2^{(s+1)}/3 – 7/3$
$(2*2^{(s+1)}+1)/3+z*4+3 < 2*2^{(s+1)})$

and hence the numbers that stand at position n+1 in the

129

considered reverse horizontal sequences can be represented as 12*r+1 and r = v*2*2^(s+1)+t, at v = 0,1,2,…., s = 2,4,6,… and t < 2*2^(s+1) or 12*r+9 and r = v*2*2^(s+1)+t, at v = 0,1,2,…., s = 3,5,7,… and t < 2*2^(s+1)

let v = 3*y + 2, t = 3*z + 0 then

$12^*(v^*2\text{^}s+t)+7 = 12^*(3^*y+2)^*2\text{^}s+12^*(3^*z+0)+7 \rightarrow$
$(12^*(3^*y^*2^*2\text{^}(s+1)+2^*2^*2\text{^}(s+1)+3^*z^*4)+4^*7 - 1)/3 =$
$= 12^*(y^*2^*2\text{^}(s+1)+(z^*4))+4^*2^*2^*2\text{^}(s+1)+9$

natural integers 2^s can be represented

$2\text{^}s = 12^*m1 + 4, s = 2,4,6,…$

or

$2\text{^}s = 12^*m2 + 8, s = 3,5,7,…$

accordingly natural integers 4*2*2*2^(s+1) can be represented

$4^*2^*2^*2\text{^}(s+1) = 12^*m3 + 8, s = 2,4,6,…$

or

$4^*2^*2^*2\text{^}(s+1) = 12^*m4 + 4, s = 3,5,7,…$

accordingly natural integers 4*2*2*2^(s+1)+5 can be represented

$4^*2^*2^*2\text{^}(s+1)+9 = 12^*m5 + 5, s = 2,4,6,…$

or

130

$4*2*2*2^{(s+1)}+9 = 12*m6 + 1, s = 3,5,7,...$

accordingly

$12*(y*2*2^{(s+1)}+(z*4))+4*2*2*2^{(s+1)}+9 =$
$12*(y*2*2^{(s+1)}+z*4+m5)+5, s = 2,4,6,...$

or

$12*(y*2*2^{(s+1)}+(z*4))+4*2*2*2^{(s+1)}+9 =$
$12*(y*2*2^{(s+1)}+z*4+m6)+1, s = 3,5,7,...$

Let us note that

$m5 = (4*2*2*2^{(s+1)}+4)/12 = (2*2*2^{(s+1)}+1)/3, s = 2,4,6,...$

$3*z < 2^s,$
$3*z < 2^{(s+1)}/2,$
$3*z =< 2^{(s+1)}/2 - 1,$
$4*z < 2*2^{(s+1)}/3 - 4/3,$
$(2*2*2^{(s+1)}+1)/3 =2*2*2^{(s+1)}/3 +1/3,$
$(2*2*2^{(s+1)}+1)/3 + 4*z < 2*2*2^{(s+1)}/3+ 2*2^{(s+1)}/3 +1/3 - 4/3$
$(2*2*2^{(s+1)}+1)/3+z*4 < 2*2^{(s+1)}$

and

$m6 = (4*2*2*2^{(s+1)}+8)/12 = (2*2*2^{(s+1)}+2)/3, s = 3,5,7,...$

$3*z < 2^s,$
$3*z < 2^{(s+1)}/2,$
$3*z =< 2^{(s+1)}/2 - 1,$
$4*z =< 2*2^{(s+1)}/3 - 4/3,$

131

$(2^{\wedge}(s+1)+2)/3 = 2^{\wedge}(s+1)/3 +2/3,$
$(2^{\wedge}(s+1)+2)/3 + 4*z =< 2*2^{\wedge}(s+1)/3+ 2*2^{\wedge}(s+1)/3 - 2/3$
$(2^{\wedge}(s+1)+2)/3+z*4 < 2*2^{\wedge}(s+1)$

and hence the numbers that stand at position n+1 in the considered reverse horizontal sequences can be represented as $12*r+5$ and $r = v*2*2^{\wedge}(s+1)+t$, at $v = 0,1,2,....,$ s = 2,4,6,… and $t < 2*2^{\wedge}(s+1)$ or $12*r+1$ and $r = v*2*2^{\wedge}(s+1)+t$, at $v = 0,1,2,....,$ s = 3,5,7,… and $t < 2*2^{\wedge}(s+1)$

let $v = 3*y + 2$, $t = 3*z + 1$ then

$12^{\star}(v^{\star}2^{\wedge}s+t)+7 = 12^{\star}(3^{\star}y+2)^{\star}2^{\wedge}s+12^{\star}(3^{\star}z+1)+7 \rightarrow$
$(12^{\star}(3^{\star}y^{\star}2^{\star}2^{\wedge}(s+1)+2^{\star}2^{\star}2^{\wedge}(s+1)+3^{\star}z^{\star}4)+12^{\star}4+4^{\star}7 - 1)/3 =$
$= 12*(y^{\star}2^{\star}2^{\wedge}(s+1)+(z^{\star}4+2))+4^{\star}2^{\star}2^{\star}2^{\wedge}(s+1)+1$

natural integers $2^{\wedge}s$ can be represented

$2^{\wedge}s = 12*m1 + 4$, s = 2,4,6,…

or

$2^{\wedge}s = 12*m2 + 8$, s = 3,5,7,…

accordingly natural integers $4*2*2*2^{\wedge}(s+1)$ can be represented

$4*2*2*2^{\wedge}(s+1) = 12*m3 + 8$, s = 2,4,6,…

or

$4*2*2*2^{\wedge}(s+1) = 12*m4 + 4$, s = 3,5,7,…

accordingly natural integers $4*2*2*2^{\wedge}(s+1)+1$ can be represented

$$4*2*2*2^{\wedge}(s+1)+1 = 12*m5 + 9, s = 2,4,6,\ldots$$

or

$$4*2*2*2^{\wedge}(s+1)+1 = 12*m6 + 5, s = 3,5,7,\ldots$$

accordingly

$$12*(y*2*2^{\wedge}(s+1)+(z*4+2))+4*2*2*2^{\wedge}(s+1)+1 =$$
$$12*(y*2*2^{\wedge}(s+1)+z*4+m5)+9, s = 2,4,6,\ldots$$

or

$$12*(y*2*2^{\wedge}(s+1)+(z*4+2))+4*2*2*2^{\wedge}(s+1)+1 =$$
$$12*(y*2*2^{\wedge}(s+1)+z*4+m6)+5, s = 3,5,7,\ldots$$

Let us note that

$$m5 = (4*2*2*2^{\wedge}(s+1)-8)/12 = (2*2*2^{\wedge}(s+1)-2)/3, s = 2,4,6,\ldots$$

$$3*z +1 < 2^{\wedge}s,$$
$$3*z +1 < 2^{\wedge}(s+1)/2,$$
$$3*z < 2^{\wedge}(s+1)/2 - 1,$$
$$4*z < 2*2^{\wedge}(s+1)/3 - 4/3,$$
$$4*z +2 < 2*2^{\wedge}(s+1)/3 +2/3,$$
$$(2*2*2^{\wedge}(s+1)-2)/3 = 2*2*2^{\wedge}(s+1)/3 -2/3,$$
$$(2*2*2^{\wedge}(s+1)-2)/3 + (4*z+2) < 2*2*2^{\wedge}(s+1)/3+$$
$$2*2^{\wedge}(s+1)/3 +2/3 - 2/3$$
$$(2*2*2^{\wedge}(s+1)-2)/3+ (z*4+2) < 2*2^{\wedge}(s+1)$$

and

$$m6 = (4*2*2*2^{\wedge}(s+1)-4)/12 = (2*2*2^{\wedge}(s+1)-1)/3, s =$$

3,5,7,...

$3*z +1 < 2^s,$
$3*z +1 < 2^{(s+1)}/2,$
$3*z =< 2^{(s+1)}/2 - 2,$
$4*z =< 2*2^{(s+1)}/3 - 8/3,$
$4*z +2 =< 2*2^{(s+1)}/3 -2/3,$
$(2*2*2 ^{(s+1)}-1)/3 = 2*2*2^{(s+1)}/3 -1/3,$
$(2*2*2^{(s+1)}-1)/3 + 4*z +2 =< 2*2*2^{(s+1)}/3+$
$2*2^{(s+1)}/3 - 1$
$(2*2*2 ^{(s+1)}-1)/3+z*4 +2 < 2*2^{(s+1)})$

and hence the numbers that stand at position n+1 in the considered reverse horizontal sequences can be represented as 12*r+9 and r = v*2*2^(s+1)+t, at v = 0,1,2,...., s = 2,4,6,... and t < 2*2^(s+1) or 12*r+5 and r = v*2*2^(s+1)+t, at v = 0,1,2,...., s = 3,5,7,... and t < 2*2^(s+1)

let v = 3*y + 2, t = 3*z + 2 then

$12^*(v^*2^s+t)+7 = 12^*(3^*y+2)^*2^s+12^*(3^*z+2)+7 \rightarrow$
$(12^*(3^*y^*2^*2^{(s+1)}+2^*2^*2^{(s+1)}+3^*z^*4)+2^*12^*4+4^*7 - 1)/3$
$=$
$= 12*(y*2*2^{(s+1)}+(z*4+3))+4*2*2*2^{(s+1)}+7$

natural integers 2^s can be represented

$2^s = 12*m1 + 4, s = 2,4,6,...$

or

$2^s = 12*m2 + 8, s = 3,5,7,...$

accordingly natural integers $4*2*2*2^{(s+1)}$ can be represented

$$4*2*2*2^{(s+1)} = 12*m3 + 8, s = 2,4,6,...$$

or

$$4*2*2*2^{(s+1)} = 12*m4 + 4, s = 3,5,7,...$$

accordingly natural integers $4*2*2*2^{(s+1)}+7$ can be represented

$$4*2*2*2^{(s+1)}+5 = 12*m5 + 1, s = 2,4,6,...$$

or

$$4*2*2*2^{(s+1)}+5 = 12*m6 + 9, s = 3,5,7,...$$

accordingly

$$12*(y*2*2^{(s+1)}+(z*4+3))+4*2*2*2^{(s+1)}+5 = 12*(y*2*2^{(s+1)}+(z*4+3)+m5)+1, s = 2,4,6,...$$

or

$$12*(y*2*2^{(s+1)}+(z*4+3))+4*2*2*2^{(s+1)}+5 = 12*(y*2*2^{(s+1)}+(z*4+3)+m6)+9, s = 3,5,7,...$$

Let us note that

$$m5 = (4*2*2*2^{(s+1)}+4)/12 = (2*2*2^{(s+1)}+1)/3, s = 2,4,6,...$$

$$3*z +2 < 2^{s},$$
$$3*z +2 < 2^{(s+1)}/2,$$
$$3*z < 2^{(s+1)}/2 - 2,$$

$3*z =< 2^{(s+1)}/2 - 3,$

$4*z =< 2*2^{(s+1)}/3 - 12/3,$

$4*z +3 =< 2*2^{(s+1)}/3 - 1,$

$(2*2*2^{(s+1)}+1)/3 = 2*2*2^{(s+1)}/3 +1/3,$

$(2*2*2^{(s+1)}+1)/3 + (4*z+2) < 2*2*2^{(s+1)}/3 + 2*2^{(s+1)}/3 +1/3 - 1$

$(2*2*2^{(s+1)}+1)/3 + (z*4+2) < 2*2^{(s+1)}$

and

$m6 = (4*2*2*2^{(s+1)}-4)/12 = (2*2*2^{(s+1)}-1)/3, s = 3,5,7,\ldots$

$3*z +2 < 2^s,$

$3*z +2 < 2^{(s+1)}/2,$

$3*z < 2^{(s+1)}/2 - 2,$

$4*z < 2*2^{(s+1)}/3 - 8/3,$

$4*z +3 < 2*2^{(s+1)}/3 +1/3,$

$(2*2*2^{(s+1)}-1)/3 = 2*2*2^{(s+1)}/3 -1/3,$

$(2*2*2^{(s+1)}-1)/3 + 4*z +3 =< 2*2*2^{(s+1)}/3 + 2*2^{(s+1)}/3$

$(2*2*2^{(s+1)}-1)/3 + z*4 +3 < 2*2^{(s+1)}$

and hence the numbers that stand at position n+1 in the considered reverse horizontal sequences can be represented as $12*r+1$ and $r = v*2*2^{(s+1)}+t$, at $v = 0,1,2,\ldots,$ $s = 2,4,6,\ldots$ and $t < 2*2^{(s+1)}$ or $12*r+9$ and $r = v*2*2^{(s+1)}+t$, at $v = 0,1,2,\ldots,$ $s = 3,5,7,\ldots$ and $t < 2*2^{(s+1)}$

13.2.3.4 Reverse horizontal sequences in which the members n, n>1 are natural integers of type 9

The reverse horizontal sequences in which the natural integers of type 9 are at position n end in these numbers because the numbers of type 9 have no images in the reverse horizontal transformation (see section 11)

Hence all reverse horizontal sequences that end in natural integers 12*p + 9, p=0,1,2,.. have length n

13.2.3.5 Reverse horizontal sequences in which the members n, n>1 are natural integers of type 11

Consider reverse horizontal sequences that begin in terms of 12*p+q and in which the terms at position n, n>1 are natural integers of type 11

$$12^*p+q \rightarrow .. \rightarrow 12^*r+11$$

let $r = v*2\wedge s+t$, where $v = 0,1,2,....$ and $t < 2\wedge s$

The following pairs of natural integers v and t are possible

$v = 3*y + 0, t = 3*z + 0$
$v = 3*y + 0, t = 3*z + 1$
$v = 3*y + 0, t = 3*z + 2$
$v = 3*y + 1, t = 3*z + 0$
$v = 3*y + 1, t = 3*z + 1$
$v = 3*y + 1, t = 3*z + 2$
$v = 3*y + 2, t = 3*z + 0$
$v = 3*y + 2, t = 3*z + 1$
$v = 3*y + 2, t = 3*z + 2$

let $v = 3*y + 0, t = 3*z + 0$ then

$$12^*(v^*2\wedge s+t)+11 = 12^*3^*y^*2\wedge s+12^*3^*z+11 \rightarrow$$
$$(12^*(3^*y^*2\wedge(s+1)+3^*z^*2)+2^*11-)/3 =$$

137

$$12*(y*2^{\wedge}(s+1)+z*2)+7$$

that is, the images of the natural integers $12*(v*2^{\wedge}s+t)+11$ in the reverse horizontal transformation will be the natural integers, $12*(y*2^{\wedge}(s+1)+z*2)+7$.

Let us note that

$3*z < 2^{\wedge}s,$
$2*z < 2^{\wedge}s,$
$2*z < 2^{\wedge}(s+1)$

and hence the numbers that stand at position n+1 in the considered reverse horizontal sequences can be represented as $12*r+7$ and $r = v*2^{\wedge}(s+1)+t$, at $v = 0,1,2,....$ and $t < 2^{\wedge}(s+1)$

let $v = 3*y + 0$, $t = 3*z + 1$ then

$$12*(v*2^{\wedge}s+t)+11 = 12*3*y*2^{\wedge}s+12*3*z+12+11 \rightarrow$$
$$(12*(3*y*2^{\wedge}(s+1)+3*z*2)+2*23 - 1)/3 =$$
$$12*(y*2^{\wedge}(s+1)+(z*2+1))+3$$

that is, the images of the natural integers $12*(v*2^{\wedge}s+t)+11$ in the reverse horizontal transformation will be the natural integers, $12*(y*2^{\wedge}(s+1)+(z*2+1))+3$.

Let us note that

$3*z + 1 < 2^{\wedge}s,$
$2*z + 1 < 2^{\wedge}s,$
$2*z + 1 < 2^{\wedge}(s+1)$

and hence the numbers that stand at position n+1 in the

considered reverse horizontal sequences can be represented as $12*r+3$ and $r = v*2^{(s+1)}+t$, at $v = 0,1,2,\ldots$ and $t < 2^{(s+1)}$

let $v = 3*y + 0$, $t = 3*z + 2$
then

$12^*(v^*2^{\wedge}s+t)+11 = 12^*3^*y^*2^{\wedge}s+12^*3^*z+12^*2+11 \rightarrow$
$(12^*(3^*y^*2^{\wedge}(s+1)+3^*z^*2)+2^*35 - 1)/3 =$
$12^*(y^*2^{\wedge}(s+1)+(z^*2+1))+11$

that is, the images of the natural integers $12*(v*2^{\wedge}s+t)+11$ in the reverse horizontal transformation will be the natural integers, $12*(y*2^{\wedge}(s+1)+(z*2+1))+11$.

Let us note that

$3*z + 2 < 2^{\wedge}s$,
$3*z + 1 < 2^{\wedge}s$,
$2*z + 1 < 2^{\wedge}s$,
$2*z + 1 < 2^{\wedge}(s+1)$

and hence the numbers that stand at position n+1 in the considered reverse horizontal sequences can be represented as $12*r+11$ and $r = v*2^{\wedge}(s+1)+t$, at $v = 0,1,2,\ldots$ and $t < 2^{\wedge}(s+1)$

let $v = 3*y + 1$, $t = 3*z + 0$ then

$12^*(v^*2^{\wedge}s+t)+11 = 12^*(3^*y+1)^*2^{\wedge}s+12^*3^*z+11 \rightarrow$
$(12^*(3^*y^*2^{\wedge}(s+1)+2^{\wedge}(s+1)+3^*z^*2)+2^*11 - 1)/3 =$
$= 12^*(y^*2^{\wedge}(s+1)+z^*2))+4^*2^{\wedge}(s+1)+7$

natural integers 2^s can be represented as

$2^s = 12*m1 + 4$, $s = 2,4,6,...$

or

$2^s = 12*m2 + 8$, $s = 3,5,7,...$

accordingly natural integers $4*2^{(s+1)}$ can be represented as

$4*2^{(s+1)} = 12*m3 + 8$, $s = 2,4,6,...$

or

$4*2^{(s+1)} = 12*m4 + 4$, $s = 3,5,7,...$

accordingly natural integers $4*2^{(s+1)}+3$ can be represented as

$4*2^{(s+1)}+7 = 12*m5 + 3$, $s = 2,4,6,...$

or

$4*2^{(s+1)}+7 = 12*m6 + 11$, $s = 3,5,7,...$

accordingly

$12*(y*2^{(s+1)}+z*2))+4*2^{(s+1)}+7 = 12*(y*2^{(s+1)}+z*2+m5)+3$, $s = 2,4,6,...$

or

$12*(y*2^{(s+1)}+z*2))+4*2^{(s+1)}+7 = 12*(y*2^{(s+1)}+z*2+m6)+11$, $s = 3,5,7,...$

Let us note that

$$m5 = (4*2^{(s+1)}+4)/12 = (2^{(s+1)}+1)/3, \ s = 2,4,6,\ldots$$

$3*z < 2^s,$
$3*z < 2^{(s+1)}/2,$
$2*z < 2^{(s+1)}/3,$
$1/3 < 2^{(s+1)}/3$
$(2^{(s+1)}+1)/3 < 2^{(s+1)}/3 + 2^{(s+1)}/3,$
$(2^{(s+1)}+1)/3 + 2*z < 2^{(s+1)}/3 + 2^{(s+1)}/2 + 2^{(s+1)}/3,$
$(2^{(s+1)}+1)/3 + z*2 < 2^{(s+1)}$

and

$$m6 = (4*2^{(s+1)}-4)/12 = (2^{(s+1)}-1)/3, \ s = 3,5,7,\ldots$$

$3*z < 2^s,$
$3*z < 2^{(s+1)}/2,$
$2*z < 2^{(s+1)}/2,$
$(2^{(s+1)}-1)/3 < 2^{(s+1)}/3,$
$(2^{(s+1)}-1)/3 + 2*z < 2^{(s+1)}/3 + 2^{(s+1)}/2$
$(2^{(s+1)}-1)/3 + z*2 < 2^{(s+1)}$

and hence the numbers that stand at position n+1 in the considered reverse horizontal sequences can be represented as $12*r+3$ and $r = v*2^{(s+1)}+t$, at $v = 0,1,2,\ldots$, $s = 2,4,6,\ldots$ and $t < 2^{(s+1)}$ or $12*r+11$ and $r = v*2^{(s+1)}+t$, at $v = 0,1,2,\ldots$, $s = 3,5,7,\ldots$ and $t < 2^{(s+1)}$

let $v = 3*y + 1$, $t = 3*z + 1$ then

$12*(v*2^s+t)+11 = 12*(3*y+1)*2^s+ 12*3*z+12+11 \rightarrow$
$(12*(3*y*2^{(s+1)}+2^{(s+1)}+3*z*2)+2*23 - 1)/3 =$
$12*(y*2^{(s+1)}+(z*2+1))+4*2^{(s+1)}+3$

natural integers 2^s can be represented as

$2^s = 12*m1 + 4$, $s = 2,4,6,...$

or

$2^s = 12*m2 + 8$, $s = 3,5,7,...$

accordingly natural integers $4*2^{(s+1)}$ can be represented as

$4*2^{(s+1)} = 12*m3 + 8$, $s = 2,4,6,...$

or

$4*2^{(s+1)} = 12*m4 + 4$, $s = 3,5,7,...$

accordingly natural integers $4*2^{(s+1)}+3$ can be represented as

$4*2^{(s+1)}+3 = 12*m5 + 11$, $s = 2,4,6,...$

or

$4*2^{(s+1)}+3 = 12*m6 + 7$, $s = 3,5,7,...$

accordingly

$12*(y*2^{(s+1)}+(z*2+1))+4*2^{(s+1)}+3 =$
$12*(y*2^{(s+1)}+(z*2+1)+m5)+11$, $s = 2,4,6,...$

or

$12*(y*2^{(s+1)}+(z*2+1))+4*2^{(s+1)}+3 =$
$12*(y*2^{(s+1)}+(z*2+1)+m6)+7$, $s = 3,5,7,...$

142

Let us note that

$m5 = (4*2^{(s+1)}-8)/12 = (2^{(s+1)}-2)/3$, $s = 2,4,6,\ldots$

$3*z+1 < 2^{s}$,
$3*z < 2^{(s+1)}/2 - 1$,
$2*z < 2^{(s+1)}/3 - 1/3$,
$2*z +1 < 2^{(s+1)}/3 +2/3$,
$(2^{(s+1)}-2)/3 = 2^{(s+1)}/3-2/3$,
$(2^{(s+1)}-2)/3 + (2*z+1) < 2^{(s+1)}/3 + 2^{(s+1)}/3$
$(2^{(s+1)}-2)/3+(z*2+1) < 2^{(s+1)}$

and

$m6 = (4*2^{(s+1)}-4)/12 = (2^{(s+1)}-1)/3$, $s = 3,5,7,\ldots$

$3*z+1 < 2^{s}$,
$3*z =< 2^{(s+1)}/2 - 2$,
$2*z =< 2^{(s+1)}/3 - 4/3$,
$2*z +1=< 2^{(s+1)}/3 -1/3$,
$(2^{(s+1)}-1)/3 = 2^{(s+1)}/3 -1/3$,
$(2^{(s+1)}-1)/3 + (2*z+1) =< 2^{(s+1)}/3 -1/3+ 2^{(s+1)}/3 - 1/3$
$(2^{(s+1)}-1)/3+ (z*2+1) < 2^{(s+1)}$

and hence the numbers that stand at position n+1 in the considered reverse horizontal sequences can be represented as $12*r+11$ and $r = v*2^{(s+1)}+t$, at $v = 0,1,2,\ldots$, $s = 2,4,6,\ldots$ and $t < 2^{(s+1)}$ or $12*r+7$ and $r = v*2^{(s+1)}+t$, at $v = 0,1,2,\ldots$, $s = 3,5,7,\ldots$ and $t < 2^{(s+1)}$

let $v = 3*y + 1$, $t = 3*z + 2$ then

$12^{*}(v^{*}2^{\wedge}s+t)+11 = 12^{*}(3^{*}y+1)^{*}2^{\wedge}s + 12^{*}3^{*}z+2^{*}12+11 \rightarrow =$
$(12^{*}(3^{*}y^{*}2^{\wedge}(s+1)+3^{*}z^{*}2)+12^{*}2^{\wedge}(s+1)+2^{*}35 - 1)/3 =$

143

$$12*(y*2^{\wedge}(s+1)+(z*2+1))+4*2^{\wedge}(s+1)+11$$

natural integers $2^{\wedge}s$ can be represented as

$$2^{\wedge}s = 12*m1 + 4, s = 2,4,6,\ldots$$

or

$$2^{\wedge}s = 12*m2 + 8, s = 3,5,7,\ldots$$

accordingly natural integers $4*2^{\wedge}(s+1)$ can be represented as

$$4*2^{\wedge}(s+1) = 12*m3 + 8, s = 2,4,6,\ldots$$

or

$$4*2^{\wedge}(s+1) = 12*m4 + 4, s = 3,5,7,\ldots$$

accordingly natural integers $4*2^{\wedge}(s+1)+11$ can be represented as

$$4*2^{\wedge}(s+1)+11 = 12*m5 + 7, s = 2,4,6,\ldots$$

or

$$4*2^{\wedge}(s+1)+11 = 12*m6 + 3, s = 3,5,7,\ldots$$

accordingly

$$12*(y*2^{\wedge}(s+1)+(z*2+1))+4*2^{\wedge}(s+1)+11 =$$
$$12*(y*2^{\wedge}(s+1)+(z*2+1)+m5)+7, s = 2,4,6,\ldots$$

or

$$12*(y*2^{\wedge}(s+1)+(z*2+1))+4*2^{\wedge}(s+1)+11 =$$

$12*(y*2^{(s+1)}+(z*2+1)+m6)+3$, $s = 3,5,7,\ldots$

Let us note that

$m5 = (4*2^{(s+1)}+4)/12 = (2^{(s+1)}+1)/3$, $s = 2,4,6,\ldots$

$3*z+2 < 2^s$,
$3*z < 2^{(s+1)}/2 - 2$,
$2*z < 2^{(s+1)}/3 - 4/3$,
$2*z +1 < 2^{(s+1)}/3 - 1/3$,
$(2^{(s+1)}+1)/3 = 2^{(s+1)}/3 + 1/3$,
$(2^{(s+1)}+1)/3 + (2*z +1) < 2^{(s+1)}/3 + 1/3 + 2^{(s+1)}/3 - 1/3$
$(2^{(s+1)}+1)/3+(z*2+1) < 2^{(s+1)}$

and

$m6 = (4*2^{(s+1)}+8)/12 = (2^{(s+1)}+2)/3$, $s = 3,5,7,\ldots$

$3*z+2 < 2^s$,
$3*z =< 2^s - 3$,
$2*z < 2^{(s+1)}/3 - 2$,
$2*z +1 < 2^{(s+1)}/3 - 1$,
$(2^{(s+1)}+2)/3 = 2^{(s+1)}/3 + 2/3$,
$(2^{(s+1)}+2)/3 + (2*z+1) < 2^{(s+1)}/3 + 2^{(s+1)}/3 - 1/3$
$(2^{(s+1)}+2)/3+(z*2+1) < 2^{(s+1)}$

and hence the numbers that stand at position n+1 in the considered reverse horizontal sequences can be represented as $12*r+7$ and $r = v*2^{(s+1)}+t$, at $v = 0,1,2,\ldots$, $s = 2,4,6,\ldots$ and $t < 2^{(s+1)}$ or $12*r+3$ and $r = v*2^{(s+1)}+t$, at $v = 0,1,2,\ldots$, $s = 3,5,7,\ldots$ and $t < 2^{(s+1)}$

let $v = 3*y + 2$, $t = 3*z + 0$ then

$12*(v*2^s+t)+11 = 12*(3*y+2)*2^s+12*3*z+11 \rightarrow$
$(12*(3*y*2^{(s+1)}+2*2^{(s+1)}+3*z*2)+2*11 - 1)/3 =$

$= 12*(y*2^{(s+1)})+z*2)+4*2*2^{(s+1)}+7$

natural integers 2^s can be represented as

$2^s = 12*m1 + 4$, $s = 2,4,6,\ldots$

or

$2^s = 12*m2 + 8$, $s = 3,5,7,\ldots$

accordingly natural integers $4*2*2^{(s+1)}$ can be represented as

$4*2*2^{(s+1)} = 12*m3 + 4$, $s = 2,4,6,\ldots$

or

$4*2*2^{(s+1)} = 12*m4 + 8$, $s = 3,5,7,\ldots$

accordingly natural integers $4*2*2^{(s+1)}+7$ can be represented as

$4*2*2^{(s+1)}+7 = 12*m5 + 11$, $s = 2,4,6,\ldots$

or

$4*2*2^{(s+1)}+7 = 12*m6 + 3$, $s = 3,5,7,\ldots$

accordingly

$12*(y*2^{(s+1)}+z*2))+4*2*2^{(s+1)}+7 =$
$12*(y*2^{(s+1)}+z*2+m5)+11$, $s = 2,4,6,\ldots$

146

or

$$12*(y*2^{\wedge}(s+1)+z*2))+4*2*2^{\wedge}(s+1)+7 =$$
$$12*(y*2^{\wedge}(s+1)+z*2+m6)+3, s = 3,5,7,...$$

Let us note that

$$m5 = (4*2*2^{\wedge}(s+1)-4)/12 = (2*2^{\wedge}(s+1)-1)/3, s = 2,4,6,...$$

$3*z < 2^{\wedge}s,$
$3*z < 2^{\wedge}(s+1)/2,$
$2*z < 2^{\wedge}(s+1)/3,$
$(2*2^{\wedge}(s+1))/3 = (2*2^{\wedge}(s+1))/3,$
$(2*2^{\wedge}(s+1)-1)/3 < (2*2^{\wedge}(s+1))/3,$
$(2*2^{\wedge}(s+1)-1)/3+z*2 < (2*2^{\wedge}(s+1))/3 + 2^{\wedge}(s+1)/3,$
$(2*2^{\wedge}(s+1)-1)/3+z*2 < 2^{\wedge}(s+1)$

and

$$m6 = (4*2*2^{\wedge}(s+1)+4)/12 = (2*2^{\wedge}(s+1)+1)/3, s = 3,5,7,...$$

$3*z < 2^{\wedge}s,$
$3*z < 2^{\wedge}(s+1)/2,$
$3*z =< 2^{\wedge}(s+1)/2-1,$
$2*z =< 2^{\wedge}(s+1)/3 - 2/3,$
$(2*2^{\wedge}(s+1))/3 = (2*2^{\wedge}(s+1)+1/3,$
$(2*2^{\wedge}(s+1)+1)/3+z*2 =< (2*2^{\wedge}(s+1))/3 + 2^{\wedge}(s+1)/3 - 1/3,$
$(2*2^{\wedge}(s+1)+1)/3+z*2 < 2^{\wedge}(s+1)$

and hence the numbers that stand at position n+1 in the considered reverse horizontal sequences can be represented as 12*r+11 and r = v*2^(s+1)+t, at v = 0,1,2,...., s = 2,4,6,... and t < 2^(s+1) or 12*r+3 and r = v*2^(s+1)+t, at v = 0,1,2,...., s = 3,5,7,... and t < 2^(s+1)

let $v = 3*y + 2$, $t = 3*z + 1$ then

$12*(v*2^s+t)+11 = 12*3*y*2^s+12*2*2^s+ 12*3*z+12+11 \rightarrow$
$(12*(3*y*2^{(s+1)}+3*z*2)+12*2*2^{(s+1)}+2*23 - 1)/3 =$
$12*(y*2^{(s+1)}+z*2)+4*2*2^{(s+1)} +3$

natural integers 2^s can be represented as

$2^s = 12*m1 + 4$, $s = 2,4,6,...$

or

$2^s = 12*m2 + 8$, $s = 3,5,7,...$

accordingly natural integers $4*2*2^{(s+1)}$ can be represented as

$4*2*2^{(s+1)} = 12*m3 + 4$, $s = 2,4,6,...$

or

$4*2*2^{(s+1)} = 12*m4 + 8$, $s = 3,5,7,...$

accordingly natural integers $4*2^{(s+1)}+3$ can be represented as

$4*2*2^s +3 = 12*m5 + 7$, $s = 2,4,6,...$

or

$4*2*2^s +3 = 12*m6 + 11$, $s = 3,5,7,...$

accordingly

$12*(y*2^{(s+1)}+z*2))+4*2*2^s +3 =$

148

$12*(y*2^{(s+1)}+z*2+m5)+7$, s = 2,4,6,…

or

$12*(y*2^{(s+1)}+z*2))+4*2*2^s +3 =$
$12*(y*2^{(s+1)}+z*2+m6)+11$, s = 3,5,7,…

Let us note that

$m5 = (4*2*2^{(s+1)}-4)/12 = (2*2^{(s+1)}-1)/3$, s = 2,4,6,…

$3*z+1 < 2^s$,
$3*z +1 < 2^{(s+1)}/2$,
$3*z < 2^{(s+1)}/2 - 1$,
$2*z < 2^{(s+1)}/3 – 2/3$,
$(2*2^{(s+1)}-1)/3 = (2*2^{(s+1)})/3-1/3$,
$(2*2^{(s+1)}-1)/3+z*2 < (2*2^{(s+1)})/3-1/3 + 2^{(s+1)}/3 – 2/3$,
$(2*2^{(s+1)}-1)/3+z*2 < 2^{(s+1)}$

and

$m6 = (4*2*2^{(s+1)}-8)/12 = (2*2^{(s+1)}-2)/3$, s = 3,5,7,…

$3*z+1 < 2^s$,
$3*z +1 < 2^{(s+1)}/2$,
$3*z < 2^{(s+1)}/2 - 1$,
$2*z < 2^{(s+1)}/3 – 2/3$,
$(2*2^{(s+1)}-2)/3 < (2*2^{(s+1)})/3-2/3$,
$(2*2^{(s+1)}-2)/3+z*2 < (2*2^{(s+1)})/3 -2/3+ 2^{(s+1)}/3 – 2/3$,
$(2*2^{(s+1)}-2)/3+z*2 < 2^{(s+1)}$

and hence the numbers that stand at position n+1 in the
considered reverse horizontal sequences can be represented

149

as $12*r+7$ and $r = v*2^{(s+1)}+t$, at $v = 0,1,2,...,$ $s = 2,4,6,...$
and $t < 2^{(s+1)}$ or $12*r+11$ and $r = v*2^{(s+1)}+t$, at $v = 0,1,2,...,$ $s = 3,5,7,...$ and $t < 2^{(s+1)}$

let $v = 3*y + 2$, $t = 3*z + 2$ then

$12*(v*2^s+t)+11 = 12*(3*y+2)*2^s+ 12*3*z+2*12+11 \rightarrow =$
$(12*(3*y*2^{(s+1)}+3*z*2)+12*2*2^{(s+1)}+2*35 - 1)/3 =$
$12*(y*2^{(s+1)}+(z*2+1))+4*2*2^{(s+1)}+11$

natural integers 2^s can be represented as

$2^s = 12*m1 + 4$, $s = 2,4,6,...$

or

$2^s = 12*m2 + 8$, $s = 3,5,7,...$

accordingly natural integers $4*2*2^{(s+1)}$ can be represented as

$4*2*2^{(s+1)} = 12*m3 + 4$, $s = 2,4,6,...$

or

$4*2*2^{(s+1)} = 12*m4 + 8$, $s = 3,5,7,...$

accordingly natural integers $4*2^{(s+1)}+11$ can be represented as

$4*2*2^s + 11 = 12*m5 + 3$, $s = 2,4,6,...$

or

$4*2*2^s + 11 = 12*m6 + 7$, $s = 3,5,7,...$

accordingly

$12*(y*2^{(s+1)}+z*2+1)+4*2*2^s +11 =$
$12*(y*2^{(s+1)}+(z*2+1)+m5)+3$, $s = 2,4,6,...$

or

$12*(y*2^{(s+1)}+z*2+1)+4*2*2^s +11 =$
$12*(y*2^{(s+1)}+(z*2+1)+m6)+7$, $s = 3,5,7,...$

Let us note that

$m5 = (4*2*2^{(s+1)}+8)/12 = (2*2^{(s+1)}+2)/3$, $s = 2,4,6,...$

$3*z+2 < 2^s$,
$3*z =< 2^{(s+1)}/2 - 3$,
$2*z =< 2^{(s+1)}/3 - 2$,
$2*z +1 =< 2^{(s+1)}/3 - 1$,
$(2*2^{(s+1)} +2)/3 = (2*2^{(s+1)})/3 +2/3$,
$(2*2^{(s+1)}+2)/3+(z*2+1) =< (2*2^{(s+1)})/3 + 2^{(s+1)}/3 - 1/3$,
$(2*2^{(s+1)}+2)/3+(z*2+1) < 2^{(s+1)}$

and

$m6 = (4*2*2^{(s+1)}+4)/12 = (2*2^{(s+1)}+1)/3$, $s = 3,5,7,...$

$3*z+2 < 2^s$,
$3*z < 2^{(s+1)}/2 - 2$,
$2*z < 2^{(s+1)}/3 - 4/3$,
$2*z +1 < 2^{(s+1)}/3 - 1/3$,
$(2*2^{(s+1)}+1)/3 = 2*2^{(s+1)}/3+1/3$,
$(2*2^{(s+1)}+1)/3+(z*2+1) < (2*2^{(s+1)})/3+ 1/3 + 2^{(s+1)}/3$

– 1/3,

$(2*2^{\wedge}(s+1)+1)/3+(z*2+1) < 2^{\wedge}(s+1)$

and hence the numbers that stand at position n+1 in the considered reverse horizontal sequences can be represented as $12*r+3$ and $r = v*2^{\wedge}(s+1)+t$, at $v = 0,1,2,....$, $s = 2,4,6,...$ and $t < 2^{\wedge}(s+1)$ or $12*r+7$ and $r = v*2^{\wedge}(s+1)+t$, at $v = 0,1,2,....$, $s = 3,5,7,...$ and $t < 2^{\wedge}(s+1)$

13.2.3.6 Reverse horizontal sequences in which the members n, n>1 are natural integers of type 1

Consider reverse horizontal sequences that begin in terms of $12*p+q$ and in which the terms at position n, n>1 are natural integers of type 1

$$12^{*}p+q \rightarrow .. \rightarrow 12^{*}r+1$$

let $r = v*2^{\wedge}s+t$, where $v = 0,1,2,....$, $t < 2^{\wedge}s$ and $r > 0$

The following pairs of natural integers v and t are possible

$v = 3*y + 0, t = 3*z + 0$
$v = 3*y + 0, t = 3*z + 1$
$v = 3*y + 0, t = 3*z + 2$
$v = 3*y + 1, t = 3*z + 0$
$v = 3*y + 1, t = 3*z + 1$
$v = 3*y + 1, t = 3*z + 2$
$v = 3*y + 2, t = 3*z + 0$
$v = 3*y + 2, t = 3*z + 1$
$v = 3*y + 2, t = 3*z + 2$

let $v = 3*y + 0, t = 3*z + 0$ then

$12^*(v^*2^s+t)+1 = 12^*(3^*y+0)^*2^s+12^*(3^*z+0)+1 \rightarrow$
$(12^*(3^*y^*4^*2^s+3^*z^*4)+4^*1 - 1)/3 =$
$12^*(y^*2^*2^\wedge(s+1)+z^*4)+1$

Let us note that

$3^*z < 2^s,$
$2^*z < 2^s,$
$2^*2^*z < 2^\wedge(s+1),$
$4^*z < 2^\wedge(s+1)$
$4^*z < 2^*2^\wedge(s+1)$

and hence the numbers that stand at position n+1 in the considered reverse horizontal sequences can be represented as 12^*r+1 and $r = v^*2^*2^\wedge(s+1)+t$, at $v = 0,1,2,\ldots$, $t < 2^*2^\wedge(s+1)$ and $r > 0$

let $v = 3^*y + 0$, $t = 3^*z + 1$ then

$12^*(v^*2^s+t)+1 = 12^*(3^*y+0)^*2^s+12^*(3^*z+1)+1 \rightarrow$
$(12^*(3^*y^*2^*2^\wedge(s+1)+3^*z^*4)+4^*12+4^*1 - 1)/3 =$
$12^*(y^*2^*2^\wedge(s+1)+(z^*4+1))+5$

Let us note that

$3^*z+1 < 2^s,$
$2^*z+1 < 2^s,$
$2^*2^*z +2 < 2^\wedge(s+1),$
$4^*z+1 < 2^\wedge(s+1)$
$4^*z+1 < 2^*2^\wedge(s+1)$

and hence the numbers that stand at position n+1 in the considered reverse horizontal sequences can be represented as 12^*r+5 and $r = v^*2^*2^\wedge(s+1)+t$, at $v = 0,1,2,\ldots$, $t <$

$2*2^{\wedge}(s+1)$ and $r > 0$

let $v = 3*y + 0$, $t = 3*z + 2$ then

$12^*(v^*2^{\wedge}s+t)+1 = 12^*(3^*y+0)^*2^{\wedge}s+12^*(3^*z+2)+1 \rightarrow$
$(12^*(3^*y^*2^*2^{\wedge}(s+1)+3^*z^*4)+4^*12^*2+4^*1 - 1)/3 =$
$= 12*(y*2*2^{\wedge}(s+1)+(z*4+2))+9$

Let us note that

$3*z+2 < 2^{\wedge}s$,
$2*z+1 < 2^{\wedge}s$,
$2*2*z+2 < 2^{\wedge}(s+1)$,

and hence the numbers that stand at position $n+1$ in the considered reverse horizontal sequences can be represented as $12*r+9$ and $r = v*2*2^{\wedge}(s+1)+t$, at $v = 0,1,2,....$, $t < 2*2^{\wedge}(s+1)$ and $r > 0$

let $v = 3*y + 1$, $t = 3*z + 0$ then

$12^*(v^*2^{\wedge}s+t)+1 = 12^*(3^*y+1)^*2^{\wedge}s+12^*3^*z+1 \rightarrow$
$(12^*(3^*y^*2^*2^{\wedge}(s+1)+2^*2^{\wedge}(s+1)+3^*z^*4)+4^*1 - 1)/3 =$
$= 12*(y*2*2^{\wedge}(s+1)+z*4))+4*2*2^{\wedge}(s+1)+1$

natural integers $2^{\wedge}s$ can be represented

$2^{\wedge}s = 12*m1 + 4$, $s = 2,4,6,...$

or

$2^{\wedge}s = 12*m2 + 8$, $s = 3,5,7,...$

accordingly natural integers $4*2*2^{\wedge}(s+1)$ can be represented

$$4*2*2^{\wedge}(s+1) = 12*m3 + 4, \ s = 2,4,6,\ldots$$

or

$$4*2*2^{\wedge}(s+1) = 12*m4 + 8, \ s = 3,5,7,\ldots$$

accordingly natural integers $4*2*2^{\wedge}(s+1)+1$ can be represented

$$4*2*2^{\wedge}(s+1)+1 = 12*m5 + 5, \ s = 2,4,6,\ldots$$

or

$$4*2*2^{\wedge}(s+1)+1 = 12*m6 + 9, \ s = 3,5,7,\ldots$$

accordingly

$$12*(y*2*2^{\wedge}(s+1)+z*4))+4*2*2^{\wedge}(s+1)+1 = 12*(y*2*2^{\wedge}(s+1)+z*4+m5)+5, \ s = 2,4,6,\ldots$$

or

$$12*(y*2*2^{\wedge}(s+1)+z*4))+4*2*2^{\wedge}(s+1)+1 = 12*(y*2*2^{\wedge}(s+1)+z*4+m6)+9, \ s = 3,5,7,\ldots$$

Let us note that

$$m5 = (4*2*2^{\wedge}(s+1)-4)/12 = (2*2^{\wedge}(s+1)-1)/3, \ s = 2,4,6,\ldots$$

$3*z < 2^{\wedge}s,$
$3*z < 2^{\wedge}(s+1)/2,$
$4*z < 2*2^{\wedge}(s+1)/3,$
$(2*2^{\wedge}(s+1)-1)/3 = 2*2^{\wedge}(s+1)/3 \ -1/3,$

155

$(2*2^\wedge(s+1)-1)/3 + 4*z < 4*2^\wedge(s+1)/3 \ {-}1/3$

$(2*2^\wedge(s+1)-1)/3+z*4 < 4*2^\wedge(s+1)/3$

$(2*2^\wedge(s+1)-1)/3+z*4 < 2*2^\wedge(s+1)$

and

$m6 = (4*2*2^\wedge(s+1)-8)/12 = (2*2^\wedge(s+1)-2)/3,\ s = 3,5,7,\ldots$

$3*z < 2^\wedge s,$

$3*z < 2^\wedge(s+1)/2,$

$4*z < 2*2^\wedge(s+1)/3,$

$(2*2^\wedge(s+1)-2)/3 = 2*2^\wedge(s+1)/3 \ {-}2/3,$

$(2*2^\wedge(s+1)-2)/3 + 4*z < 4*2^\wedge(s+1)/3 \ {-}2/3$

$(2*2^\wedge(s+1)-2)/3+z*4 < 4*2^\wedge(s+1)/3$

$(2*2^\wedge(s+1)-2)/3+z*4 < 2*2^\wedge(s+1)$

and hence the numbers that stand at position n+1 in the considered reverse horizontal sequences can be represented as $12*r+5$ and $r = v*2*2^\wedge(s+1)+t$, at $v = 0,1,2,\ldots$, $s = 2,4,6,\ldots$ and $t < 2*2^\wedge(s+1)$ or $12*r+9$ and $r = v*2*2^\wedge(s+1)+t$, at $v = 0,1,2,\ldots$, $s = 3,5,7,\ldots$ and $t < 2*2^\wedge(s+1)$

let $v = 3*y + 1$, $t = 3*z + 1$ then

$12*(v*2^\wedge s+t)+1 = 12*(3*y+1)*2^\wedge s+12*(3*z+1)+1 \rightarrow$
$(12*(3*y*2*2^\wedge(s+1)+2*2^\wedge(s+1)+3*z*4)+4*12+4*1 - 1)/3 =$
$12*(y*2*2^\wedge(s+1)+(z*4+1))+4*2*2^\wedge(s+1)+5$

natural integers $2^\wedge s$ can be represented

$2^\wedge s = 12*m1 + 4,\ s = 2,4,6,\ldots$

or

$2^\wedge s = 12*m2 + 8,\ s = 3,5,7,\ldots$

accordingly natural integers $4*2*2^{(s+1)}$ can be represented

$$4*2*2^{(s+1)} = 12*m3 + 4, s = 2,4,6,...$$

or

$$4*2*2^{(s+1)} = 12*m4 + 8, s = 3,5,7,...$$

accordingly natural integers $4*2*2^{(s+1)}+5$ can be represented

$$4*2*2^{(s+1)}+5 = 12*m5 + 9, s = 2,4,6,...$$

or

$$4*2*2^{(s+1)}+5 = 12*m6 + 1, s = 3,5,7,...$$

accordingly

$$12*(y*2*2^{(s+1)}+(z*4+1))+4*2^{(s+1)}+5 = 12*(y*2^{(s+1)}+(z*4+1)+m5)+9, s = 2,4,6,...$$

or

$$12*(y*2*2^{(s+1)}+(z*4+1))+4*2^{(s+1)}+5 = 12*(y*2^{(s+1)}+(z*4+1)+m6)+1, s = 3,5,7,...$$

Let us note that

$$m5 = (4*2^{(s+1)}-4)/12 = (2*2^{(s+1)}-1)/3, s = 2,4,6,...$$

$$3*z +1 < 2^{s},$$
$$3*z + 1 < 2^{(s+1)}/2,$$
$$3*z < 2^{(s+1)}/2 - 1,$$

$3*z =< 2^{\wedge}(s+1)/2 — 2,$

$4*z =< 2*2^{\wedge}(s+1)/3 - 2/3,$

$4*z + 1 =< 2*2^{\wedge}(s+1)/3 +1/3,$

$(2*2^{\wedge}(s+1)-1)/3 = 2*2^{\wedge}(s+1)/3 - 1/3,$

$(2*2^{\wedge}(s+1)-1)/3 + 4*z +1 =< 2*2^{\wedge}(s+1)/3+ 2*2^{\wedge}(s+1)/3 - 1$

$(2*2^{\wedge}(s+1)-1)/3+z*4 +1< 2*2^{\wedge}(s+1)$

and

$m6 = (4*2^{\wedge}(s+1)+4)/12 = (2*2^{\wedge}(s+1)+1)/3, \ s = 3,5,7,...$

$3*z +1< 2^{\wedge}s,$

$3*z + 1< 2^{\wedge}(s+1)/2,$

$3*z < 2^{\wedge}(s+1)/2 — 1,$

$4*z < 2*2^{\wedge}(s+1)/3 - 1/3,$

$(2*2^{\wedge}(s+1)+1)/3 = 2^{\wedge}(s+1)/3 +1/3,$

$(2*2^{\wedge}(s+1)+1)/3 + 4*z+1 < 2*2^{\wedge}(s+1)/3+ 2*2^{\wedge}(s+1)/3$

$(2*2^{\wedge}(s+1)+1)/3+z*4+1< 2*2^{\wedge}(s+1)$

and hence the numbers that stand at position n+1 in the considered reverse horizontal sequences can be represented as $12*r+9$ and $r = v*2*2^{\wedge}(s+1)+t$, at v = 0,1,2,...., s = 2,4,6,... and $t < 2*2^{\wedge}(s+1)$ or $12*r+1$ and $r = v*2*2^{\wedge}(s+1)+t$, at v = 0,1,2,...., s = 3,5,7,... and $t < 2*2^{\wedge}(s+1)$

let $v = 3*y + 1$, $t = 3*z + 2$ then

$12^{*}(v*2^{\wedge}s+t)+1 = 12^{*}(3^{*}y+1)^{*}2^{\wedge}s+12^{*}(3^{*}z+2)+1 \rightarrow$

$(12^{*}(3^{*}y*2*2^{\wedge}(s+1)+2*2^{\wedge}(s+1)+3^{*}z*4)+4*12*2+4*1 – 1)/3 =$

$= 12*(y*2*2^{\wedge}(s+1)+(z*4+2))+4*2*2^{\wedge}(s+1)+9$

natural integers $2^{\wedge}s$ can be represented

$2^\wedge s = 12*m1 + 4, s = 2,4,6,\ldots$

or

$2^\wedge s = 12*m2 + 8, s = 3,5,7,\ldots$

accordingly natural integers $4*2*2^\wedge(s+1)$ can be represented

$4*2^\wedge(s+1) = 12*m3 + 4, s = 2,4,6,\ldots$

or

$4*2^\wedge(s+1) = 12*m4 + 8, s = 3,5,7,\ldots$

accordingly natural integers $2*2^\wedge(s+1)+9$ can be represented

$4*2*2^\wedge(s+1)+9 = 12*m5 + 1, s = 2,4,6,\ldots$

or

$4*2*2^\wedge(s+1)+9 = 12*m6 + 5, s = 3,5,7,\ldots$

accordingly

$12*(y*2*2^\wedge(s+1)+(z*4+2))+4*2*2^\wedge(s+1)+9 = 12*(y*2*2^\wedge(s+1)+(z*4+2)+m5)+1, s = 2,4,6,\ldots$

or

$12*(y*2*2^\wedge(s+1)+(z*4+2))+4*2*2^\wedge(s+1)+9 = 12*(y*2*2^\wedge(s+1)+(z*4+2)+m6)+5, s = 3,5,7,\ldots$

Let us note that

$m5 = (2*2^{(s+1)}+8)/12 = (2*2^{(s+1)}-1)/3, s = 2,4,6,...$

$3*z +2< 2^s,$
$3*z + 2< 2^{(s+1)}/2,$
$3*z < 2^{(s+1)}/2 - 2,$
$4*z < 2*2^{(s+1)}/3 - 8/3,$
$4*z +2< 2*2^{(s+1)}/3 -2/3,$
$(2*2^{(s+1)}+2)/3 = 2*2^{(s+1)}/3 +2/3,$
$(2*2^{(s+1)}+2)/3 + 4*z+2 < 2*2^{(s+1)}/3+ 2*2^{(s+1)}/3$
$(2*2^{(s+1)}+2)/3+z*4 +2 < 2*2^{(s+1)}$

and

$m6 = (2*2^{(s+1)}+4)/12 = (2*2^{(s+1)}+1)/3, s = 3,5,7,...$

$3*z +2< 2^s,$
$3*z + 2< 2^{(s+1)}/2,$
$3*z < 2^{(s+1)}/2 - 2,$
$4*z < 2*2^{(s+1)}/3 - 8/3,$
$4*z +2< 2*2^{(s+1)}/3 -2/3,$
$(2*2^{(s+1)}+1)/3 = 2*2^{(s+1)}/3 +1/3,$
$(2*2^{(s+1)}+1)/3 + 4*z+2 < 2*2^{(s+1)}/3+ 2*2^{(s+1)}/3 - 1/3$
$(2*2^{(s+1)}+1)/3+z*4 +2 < 2*2^{(s+1)}$

and hence the numbers that stand at position n+1 in the
considered reverse horizontal sequences can be represented
as $12*r+1$ and $r = v*2*2^{(s+1)}+t$, at $v = 0,1,2,...., s = 2,4,6,...$ and $t < 2*2^{(s+1)}$ or $12*r+5$ and $r = v*2*2^{(s+1)}+t$, at $v = 0,1,2,...., s = 3,5,7,...$ and $t < 2*2^{(s+1)}$

let $v = 3*y + 2, t = 3*z + 0$ then

$12*(v*2^s+t)+1 = 12*(3*y+2)*2^s+12*(3*z+0)+1 \rightarrow$
$(12*(3*y*2*2^{(s+1)}+2*2*2^{(s+1)}+3*z*4)+4*1 - 1)/3 =$
$12*(y*2*2^{(s+1)}+(z*4))+4*2*2*2^{(s+1)}+1$

natural integers 2^s can be represented

$2^s = 12*m1 + 4$, $s = 2,4,6,...$

or

$2^s = 12*m2 + 8$, $s = 3,5,7,...$

accordingly natural integers $4*2*2*2^{(s+1)}$ can be represented

$4*2*2*2^{(s+1)} = 12*m3 + 8$, $s = 2,4,6,...$

or

$4*2*2*2^{(s+1)} = 12*m4 + 4$, $s = 3,5,7,...$

accordingly natural integers $4*2*2*2^{(s+1)}+1$ can be represented

$4*2*2*2^{(s+1)}+1 = 12*m5 + 9$, $s = 2,4,6,...$

or

$4*2*2*2^{(s+1)}+1 = 12*m6 + 5$, $s = 3,5,7,...$

accordingly

$12*(y*2*2^{(s+1)}+(z*4))+4*2*2*2^{(s+1)}+1 =$
$12*(y*2*2^{(s+1)}+z*4+m5)+9$, $s = 2,4,6,...$

or

$12*(y*2*2^{\wedge}(s+1)+(z*4))+4*2*2*2^{\wedge}(s+1)+1 =$
$12*(y*2*2^{\wedge}(s+1)+z*4+m6)+5, s = 3,5,7,...$

Let us note that

$m5 = (4*2*2*2^{\wedge}(s+1)-8)/12 = (2*2*2^{\wedge}(s+1)-2)/3, s = 2,4,6,...$

$3*z < 2^{\wedge}s,$
$3*z < 2^{\wedge}(s+1)/2,$
$4*z < 2*2^{\wedge}(s+1)/3,$
$(2*2*2^{\wedge}(s+1)-2)/3 = 2*2*2^{\wedge}(s+1)/3 -2/3,$
$(2*2*2^{\wedge}(s+1)-2)/3 + 4*z < 2*2*2^{\wedge}(s+1)/3+ 2*2^{\wedge}(s+1)/3 - 2/3$
$(2*2*2^{\wedge}(s+1)-2)/3+z*4 < 2*2^{\wedge}(s+1)$

and

$m6 = (4*2*2*2^{\wedge}(s+1)-4)/12 = (2*2*2^{\wedge}(s+1)-1)/3, s = 3,5,7,...$

$3*z < 2^{\wedge}s,$
$3*z < 2^{\wedge}(s+1)/2,$
$4*z < 2*2^{\wedge}(s+1)/3,$
$(2*2*2^{\wedge}(s+1)-1)/3 = 2*2*2^{\wedge}(s+1)/3 -1/3,$
$(2*2*2^{\wedge}(s+1)-1)/3 + 4*z < 2*2*2^{\wedge}(s+1)/3+ 2*2^{\wedge}(s+1)/3 - 1/3$
$(2*2*2^{\wedge}(s+1)-1)/3+z*4 < 2*2^{\wedge}(s+1)$

and hence the numbers that stand at position n+1 in the considered reverse horizontal sequences can be represented as $12*r+9$ and $r = v*2*2^{\wedge}(s+1)+t$, at $v = 0,1,2,....$, $s = 2,4,6,...$ and $t < 2*2^{\wedge}(s+1)$ or $12*r+5$ and $r = v*2*2^{\wedge}(s+1)+t$, at $v = 0,1,2,....$, $s = 3,5,7,...$ and $t < 2*2^{\wedge}(s+1)$

162

let $v = 3*y + 2$, $t = 3*z + 1$ then

$12^*(v^*2\text{^}s+t)+1 = 12^*(3^*y+2)^*2\text{^}s+12^*(3^*z+1)+1 \rightarrow$
$(12^*(3^*y^*2^*2\text{^}(s+1)+2^*2^*2\text{^}(s+1)+3^*z^*4)+12^*4+4^*1 - 1)/3 =$
$= 12^*(y^*2^*2\text{^}(s+1)+(z^*4+1))+4^*2^*2^*2\text{^}(s+1)+5$

natural integers $2\text{^}s$ can be represented

$2\text{^}s = 12^*m1 + 4$, $s = 2,4,6,...$

or

$2\text{^}s = 12^*m2 + 8$, $s = 3,5,7,...$

accordingly natural integers $4^*2^*2^*2\text{^}(s+1)$ can be represented

$4^*2^*2^*2\text{^}(s+1) = 12^*m3 + 8$, $s = 2,4,6,...$

or

$4^*2^*2^*2\text{^}(s+1) = 12^*m4 + 4$, $s = 3,5,7,...$

accordingly natural integers $4^*2^*2^*2\text{^}(s+1)+5$ can be represented

$4^*2^*2^*2\text{^}(s+1)+5 = 12^*m5 + 1$, $s = 2,4,6,...$

or

$4^*2^*2^*2\text{^}(s+1)+5 = 12^*m6 + 9$, $s = 3,5,7,...$

accordingly

163

$12*(y*2*2^{(s+1)}+(z*4+2))+4*2*2*2^{(s+1)}+5 =$
$12*(y*2*2^{(s+1)}+(z*4+1)+m5)+1,\ s = 2,4,6,\ldots$

or

$12*(y*2*2^{(s+1)}+(z*4+2))+4*2*2*2^{(s+1)}+5 =$
$12*(y*2*2^{(s+1)}+(z*4+1)+m6)+9,\ s = 3,5,7,\ldots$

Let us note that

$m5 = (4*2*2*2^{(s+1)}+4)/12 = (2*2*2^{(s+1)}+1)/3,\ s = 2,4,6,\ldots$

$3*z +1 < 2^s,$
$3*z +1 < 2^{(s+1)}/2,$
$3*z < 2^{(s+1)}/2 - 1,$
$4*z < 2*2^{(s+1)}/3 - 4/3,$
$4*z +2 < 2*2^{(s+1)}/3 +2/3,$
$(2*2*2^{(s+1)}+1)/3 = 2*2*2^{(s+1)}/3 +1/3,$
$(2*2*2^{(s+1)}+1)/3 + (4*z+1) < 2*2*2^{(s+1)}/3 + 2*2^{(s+1)}/3 +2/3 - 2/3$
$(2*2*2^{(s+1)}+1)/3 + (z*4+1) < 2*2^{(s+1)}$

and

$m6 = (4*2*2*2^{(s+1)}-4)/12 = (2*2*2^{(s+1)}-1)/3,\ s = 3,5,7,\ldots$

$3*z +1 < 2^s,$
$3*z +1 < 2^{(s+1)}/2,$
$3*z < 2^{(s+1)}/2 - 1,$
$4*z < 2*2^{(s+1)}/3 - 4/3,$
$4*z +1 < 2*2^{(s+1)}/3 -1/3,$
$(2*2*2^{(s+1)}-1)/3 = 2*2*2^{(s+1)}/3 -1/3,$
$(2*2*2^{(s+1)}-1)/3 + (4*z +1) =< 2*2*2^{(s+1)}/3 + 2*2^{(s+1)}/3 - 2/3$

$(2*2*2 \char`\^(s+1)-1)/3+(z*4 +1) < 2*2\char`\^(s+1)$

and hence the numbers that stand at position n+1 in the considered reverse horizontal sequences can be represented as $12*r+1$ and $r = v*2*2\char`\^(s+1)+t$, at $v = 0,1,2,\ldots$, $s = 2,4,6,\ldots$ and $t < 2*2\char`\^(s+1)$ or $12*r+9$ and $r = v*2*2\char`\^(s+1)+t$, at $v = 0,1,2,\ldots$, $s = 3,5,7,\ldots$ and $t < 2*2\char`\^(s+1)$

let $v = 3*y + 2$, $t = 3*z + 2$ then

$12^{\star}(v^{\star}2\char`\^s+t)+1 = 12^{\star}(3^{\star}y+2)^{\star}2\char`\^s+12^{\star}(3^{\star}z+2)+1 \rightarrow$
$(12^{\star}(3^{\star}y^{\star}2^{\star}2\char`\^(s+1)+2^{\star}2^{\star}2\char`\^(s+1)+3^{\star}z^{\star}4)+2^{\star}12^{\star}4+4^{\star}1 - 1)/3$
$=$
$12*(y*2*2\char`\^(s+1)+(z*4+2))+4*2*2*2\char`\^(s+1)+9$

natural integers $2\char`\^s$ can be represented

$2\char`\^s = 12*m1 + 4$, $s = 2,4,6,\ldots$

or

$2\char`\^s = 12*m2 + 8$, $s = 3,5,7,\ldots$

accordingly natural integers $4*2*2*2\char`\^(s+1)$ can be represented

$4*2*2*2\char`\^(s+1) = 12*m3 + 8$, $s = 2,4,6,\ldots$

or

$4*2*2*2\char`\^(s+1) = 12*m4 + 4$, $s = 3,5,7,\ldots$

accordingly natural integers $4*2*2*2\char`\^(s+1)+9$ can be represented

$4*2*2*2^\wedge(s+1)+9 = 12*m5 + 5, s = 2,4,6,\ldots$

or

$4*2*2*2^\wedge(s+1)+9 = 12*m6 + 1, s = 3,5,7,\ldots$

accordingly

$12*(y*2*2^\wedge(s+1)+(z*4+2))+4*2*2*2^\wedge(s+1)+9 =$
$12*(y*2*2^\wedge(s+1)+(z*4+2)+m5)+5, s = 2,4,6,\ldots$

or

$12*(y*2*2^\wedge(s+1)+(z*4+2))+4*2*2*2^\wedge(s+1)+9 =$
$12*(y*2*2^\wedge(s+1)+(z*4+2)+m6)+1, s = 3,5,7,\ldots$

Let us note that

$m5 = (4*2*2*2^\wedge(s+1)+4)/12 = (2*2*2^\wedge(s+1)+1)/3, s =$
$2,4,6,\ldots$

$3*z +2< 2^\wedge s,$
$3*z +2< 2^\wedge(s+1)/2,$
$3*z < 2^\wedge(s+1)/2 - 2,$
$4*z < 2*2^\wedge(s+1)/3 - 8/3,$
$4*z +2 < 2*2^\wedge(s+1)/3 - 2/3,$
$(2*2*2^\wedge(s+1)+1)/3 =2*2*2^\wedge(s+1)/3 +1/3,$
$(2*2*2^\wedge(s+1)+1)/3 + (4*z+2) < 2*2*2^\wedge(s+1)/3+$
$2*2^\wedge(s+1)/3 - 1/3$
$(2*2*2^\wedge(s+1)+1)/3+ (z*4+2) < 2*2^\wedge(s+1)$

and

$m6 = (4*2*2*2^\wedge(s+1)+8)/12 = (2*2*2^\wedge(s+1)+2)/3, s =$

166

3,5,7,...

$3*z +2< 2^s$,
$3*z +2< 2^{(s+1)}/2$,
$3*z < 2^{(s+1)}/2 - 2$,
$4*z < 2*2^{(s+1)}/3 - 8/3$,
$4*z +2 < 2*2^{(s+1)}/3 - 2/3$,
$(2*2*2 ^{(s+1)}+2)/3 = 2*2*2^{(s+1)}/3 +2/3$,
$(2*2*2^{(s+1)}+2)/3 + 4*z +2 < 2*2*2^{(s+1)}/3 + 2*2^{(s+1)}/3$
$(2*2*2 ^{(s+1)}+2)/3+z*4 +2 < 2*2^{(s+1)}$

and hence the numbers that stand at position n+1 in the considered reverse horizontal sequences can be represented as $12*r+5$ and $r = v*2*2^{(s+1)}+t$, at v = 0,1,2,...., s = 2,4,6,... and $t < 2*2^{(s+1)}$ or $12*r+1$ and $r = v*2*2^{(s+1)}+t$, at v = 0,1,2,...., s = 3,5,7,... and $t < 2*2^{(s+1)}$

13.2.3.7 Consequence from sections 13.2.3.1 to 13.2.3.6

It follows from sections 13.2.3.1 to 13.2.3.6 and 12.1 that for any p > = 0 if q = 3,5,7,9,11, and p > 0 if q =1, and any m > 0 in a set of reverse horizontal sequences

$12*p+q$, $12*p+(q+2)$, $12*p+(q+4)$, $12*p+(q+6)$, $12*p+(q+8)$, $12*p+(q+10)$,
$12*(p+1)+q$, $12*(p+1)+(q+2)$, $12*(p+1)+(q+4)$, $12*(p+1)+(q+6)$,
$12*(p+1)+(q+8)$, $12*(p+1)+(q+10)$,....,$12*(p+3^{(n-1)}*3^m-1)+q$, $12*(p+3^{(n-1)}*3^m-1)+(q+2)$, $12*(p+3^{(n-1)}*3^m-1)+(q+4)$, $12*(p+3^{(n-1)}*3^m-1)+(q+6)$, $12*(p+3^{(n-1)}*3^m-1)+(q+8)$, $12*(p+3^{(n-1)}*3^m-1)+(q+10)$

exactly $(6*3^{(n-1)}*3^m)*(2/3)^n$ have length greater than n.

Hence (see section 12.2) the proportion of sequences whose length is greater than n in any set (whose size is a multiple of $6*3^{(n-1)}*3^n$) of straight horizontal sequences that are arranged in a row is $(2/3)^n$.

14. The absence of "loops" in the horizontal sequences

The following will show that all members of all horizontal sequences are different natural integers.

14.1 The absence of "loops" in the direct horizontal sequences

We show that all direct horizontal sequences do not have the same terms

Consider all straight horizontal sequences

14.1.1 Direct horizontal sequences that begin in natural integers of type 3 and 9

All direct horizontal sequence, which starts in the natural integers of type 3 or 9 does not have "loops" or that the same do not have duplicate members.
All natural integers in such sequences are different.

The fact that the natural integers of type 3 and 9 there are

168

no prototypes in the direct horizontal transformation.

The direct horizontal transformation is one-to-one and each number if has a prototype, then only one

Therefore, any term in a direct horizontal sequence that starts with a number of type 3 or 9 can be reach by applying a straight horizontal transformation F sequentially starting with a natural integer of type 3 or 9

Accordingly, from any member of a direct horizontal sequence that starts with a number of type 3 or 9, you can reach to the beginning of the sequence by applying the reverse horizontal transformation G sequentially

Therefore, in direct horizontal sequences that begin in numbers of type 3 or 9, there is no member that can be reached twice in sequence by applying a straight horizontal transformation

14.1.2 Direct horizontal sequences that begin in vertical natural integers

All direct horizontal sequences that begin in the vertical natural integers end in these vertical natural integers
Therefore, there are no loops in such sequences either

14.1.3 Direct horizontal sequences that end in vertical natural integers

All direct horizontal sequences that end in vertical numbers do not have loops for the same reason that direct horizontal sequences that start in numbers of type 3 or 9 do not have loops

Because from any member of such sequences it is possible
to reach in a finite number of steps to the last member of
such sequences in a vertical number, consistently applying
a direct horizontal transformation

Accordingly, any member of such a sequence can be
reached in a finite number of steps from the last member,
which is a vertical number, consistently applying the
reverse horizontal transformation

14.1.4 Direct horizontal sequences that end in other natural integers

Namely, sequences that do not begin in natural integers of
type 3 and 9, and that do not begin and end in vertical
natural integers

It is obvious that if a direct horizontal sequence contains a
loop then all this sequence consists only of the loop
otherwise, there would be a member of this sequence,
which would be the prototype of two natural integers

Earlier in the previous sections, we have already noted that
any odd natural integer greater than 1 can be represented as

$12*(v*3^s+t)+q$, где $v=0,1,2,3...$, $t<3^s$, $q = 3,5,7,9,11,1$

Therefore, if natural integer occurs again in a straight
horizontal sequence through n consecutive direct
transformations then the following equations must be truth

$12*(v[0]*3^s+t[0])+q = 12*(v[1]*3^{(s+n)}+t[1])+q =$
$12*(v[2]*3^{(s+2*n)}+t[2])+q = ...$

$12*(v[i]*3^{(s+i*n)}+t[i])+q = \ldots\ldots i = 0,1,2,3,\ldots.$

$12*(v[0]*3^{s}+t[0])+q = 12*(v[1]*3^{(s+n)}+t[1])+q$

$v[0]*3^{s}+t[0] = v[1]*3^{(s+n)}+t[1]$
$t[0] - t[1] = v[1]*3^{(s+n)} - v[0]*3^{s}$

$v[1]*3^{(s+n)} - v[0]*3^{s} <= t[0] < 3^{s}$

$v[1]*3^{(s+n)} - v[0]*3^{s} < 3^{s}$

$v[1]*3^{n} - v[0] < 1$

what is impossible for v !=0.

Consider direct horizontal sequences, all whose members can be represented as $12*(t)+q$, that is, when $v = 0$ for all members of the sequence

In this case $t < 3$ because otherwise it could be represented as $t = v*3^{s}+ t1$ where $s > 0$, $t1 < 3^{s}$ and $v > 0$

However, above we have shown that the natural integers which can be represented as

$12*(v*3^{s}+t)+q$ где $s > 0$, $t1 < 3^{s}$ и $v > 0$ can't be members of the loop

But for $t < 3$ all 11 direct horizontal sequences, which start in natural integers $12*(0)+5$, $12*(0)+7$, $12*(0)+11$, $12*(1)+1$, $12*(1)+5$, $12*(1)+7$, $12*(1)+11$, $12*(2)+1$, $12*(2)+5$, $12*(2)+7$, $12*(2)+11$, do not have loops and are finite.

14.2 The absence of "loops" in the reverse horizontal sequences

In order to show that reverse horizontal sequences do not contain loops, we could use arguments similar to those we used in proving the absence of loops in direct horizontal sequences

However, this is not necessary

If there was at least one reverse horizontal sequence in the form of a loop, then there would necessarily be exactly the same direct horizontal sequence, which would consist of the same natural integers, representing a direct horizontal sequence, if you apply a direct horizontal transformation to it
But in the previous paragraph we showed that direct horizontal sequences do not contain loops

15. All horizontal sequences are finite

15.1 All direct horizontal sequences are finite

15.1.1 The first proof of finiteness of direct horizontal sequences

Suppose that there is an infinite direct sequence of horizontal natural integers of type 5, 7, 11 or 1
As follows from sections 13.1.1-13.1.3, all natural integers in this sequence are different

172

Thus, each sequence that starts at any of these numbers is infinite

Therefore, the" density " of the distribution of sequences of arbitrarily large length is greater than 0

However, in previous sections 13.1.1-13.1.2 it was shown that the "density" of straight horizontal sequences whose length is greater than or equal to n is $(3/4)^n$

That is, the" density " of the distribution of such sequences tends to zero and therefore the assumption of the existence of an infinite straight horizontal sequence is incorrect

15.1.2 The second proof of finiteness of direct horizontal sequences. Reverse Structure of the Collatz and Reverse Fibonacci Structure of the Collatz

Further everywhere in section 15.1.2 the symbol → will denote the reverse horizontal transformation G and s will denote the positive integer

We further show that ALL DIRECT horizontal sequences are of finite length

Dear readers, there is no typo in this place

The following will show that the DIRECT horizontal sequences have a finite length. This prove is based on the properties of the reverse horizontal sequences

Consider the following structure, which consists of columns.

Everywhere further k=0,1,2,...

In the column with the number 0 we place vertical natural integers of type 5, 9 and type 1

12*(2*k+0)+5
12*(2*k+1)+9
12*(2*k+1)+1

the natural integers 12*(2*k+1)+9 have no images in the reverse horizontal transformation

all other natural integers have images in the reverse horizontal transformation

let's place these images in the column with the number 1

Let RHT is reverse horizontal transformation

$12^*(2^*(3^*k+0)+0)+5 \rightarrow 12^*(k^*2\char`^2+0)+3$ do not have the images at the RHT
$12^*(2^*(3^*k+1)+0)+5 \rightarrow 12^*(k^*2\char`^2+1)+7$
$12^*(2^*(3^*k+2)+0)+5 \rightarrow 12^*(k^*2\char`^2+2)+11$
12*(2*k+1)+9 do not have the images at the RHT
$12^*(2^*(3^*k+0)+1)+1 \rightarrow 12^*(k^*2\char`^3+1)+5$
$12^*(2^*(3^*k+1)+1)+1 \rightarrow 12^*(k^*2\char`^3+4)+1$
$12^*(2^*(3^*k+2)+1)+1 \rightarrow 12^*(k^*2\char`^3+6)+9$ do not have the images at the RHT

the natural integers 12*(k*2^2+0)+3 и 12*(k*2^3+6)+9 have no images in the reverse horizontal transformation

all other natural integers have images in the reverse horizontal transformation

let's place these images in the column with the number 2

$12^*((3^*k+0)^*2\char`^2+1)+7 \rightarrow 12^*(k^*2\char`^4+2)+1$
$12^*((3^*k+1)^*2\char`^2+1)+7 \rightarrow 12^*(k^*2\char`^4+7)+5$

12*((3*k+2)*2^2+1)+7 → 12*(k*2^4+12)+9 do not have the images at the RHT

12*((3*k+0)*2^2+2)+11 → 12*(k*2^3+1)+11

12*((3*k+1)*2^2+2)+11 → 12*(k*2^3+4)+ 7

12*((3*k+2)*2^2+2)+11 → 12*(k*2^3+7)+ 3 do not have the images at the RHT

12*((3*k+0)*2^3+1)+5 → 12*(k*2^4)+0)+11

12*((3*k+1)*2^3+1)+5 → 12*(k*2^4)+6)+3 do not have the images at the RHT

12*((3*k+2)*2^3+1)+5 → 12*(k*2^4)+11)+7

12*((3*k+0)*2^3+4)+1 → 12*(k*2^5+5)+5

12*((3*k+1)*2^3+4)+1 → 12*(k*2^5+16)+1

12*((3*k+2)*2^3+4)+1 → 12*(k*2^5+26)+9 do not have the images at the RHT

the natural integers 12*(k*2^4+12)+9, 12*(k*2^3+3)+3, 12*(k*2^4)+6)+3, 12*(k*2^5+26)+9 have no images in the reverse horizontal transformation

all other natural integers have images in the reverse horizontal transformation

let's place these images in the column with the number 3

12*((3*k+0)*2^4+2)+1 → 12*(k*2^6+2)+9 do not have the images at the RHT

12*((3*k+1)*2^4+2)+1 → 12*(k*2^6+24)+1

12*((3*k+2)*2^4+2)+1 → 12*(k*2^6+45)+5

12*((3*k+0)*2^4+7)+5 → 12*(k*2^5+4)+11

12*((3*k+1)*2^4+7)+5 → 12*(k*2^5+15)+7

12*((3*k+2)*2^4+7)+5 → 12*(k*2^5+26)+3 do not have the images at the RHT

$12^*((3^*k+0)^*2^\wedge 3+1)+11 \rightarrow 12^*(k^*2^\wedge 4+1)+ 3$ do not have the images at the RHT

$12^*((3^*k+1)^*2^\wedge 3+1)+11 \rightarrow 12^*(k^*2^\wedge 4+6)+ 7$

$12^*((3^*k+2)^*2^\wedge 3+1)+11 \rightarrow 12^*(k^*2^\wedge 4+11)+11$

$12^*((3^*k+0)^*2^\wedge 3+4)+7 \rightarrow 12^*(k^*2^\wedge 5+6)+ 1$

$12^*((3^*k+1)^*2^\wedge 3+4)+7 \rightarrow 12^*(k^*2^\wedge 5+16)+ 9$ do not have the images at the RHT

$12^*((3^*k+2)^*2^\wedge 3+4)+7 \rightarrow 12^*(k^*2^\wedge 5+27)+ 5$

$12^*((3^*k+0)^*2^\wedge 4+0)+11 \rightarrow 12^*(k^*2^\wedge 5+0)+ 7$

$12^*((3^*k+1)^*2^\wedge 4+0)+11 \rightarrow 12^*(k^*2^\wedge 5+11)+ 3$ do not have the images at the RHT

$12^*((3^*k+2)^*2^\wedge 4+0)+11 \rightarrow 12^*(k^*2^\wedge 5+21)+11$

$12^*((3^*k+0)^*2^\wedge 4)+11)+7 \rightarrow 12^*(k^*2^\wedge 6+15)+5$

$12^*((3^*k+1)^*2^\wedge 4)+11)+7 \rightarrow 12^*(k^*2^\wedge 6+36)+9$ do not have the images at the RHT

$12^*((3^*k+2)^*2^\wedge 4)+11)+7 \rightarrow 12^*(k^*2^\wedge 6+58)+1$

$12^*((3^*k+0)^*2^\wedge 5+5)+5 \rightarrow 12^*(k^*2^\wedge 6+3)+7$

$12^*((3^*k+1)^*2^\wedge 5+5)+5 \rightarrow 12^*(k^*2^\wedge 6+24)+11$

$12^*((3^*k+2)^*2^\wedge 5+5)+5 \rightarrow 12^*(k^*2^\wedge 6+46)+3$ do not have the images at the RHT

$12^*((3^*k+0)^*2^\wedge 5+16)+1 \rightarrow 12^*(k^*2^\wedge 7+21)+5$

$12^*((3^*k+1)^*2^\wedge 5+16)+1 \rightarrow 12^*(k^*2^\wedge 7+64)+1$

$12^*((3^*k+2)^*2^\wedge 5+16)+1 \rightarrow 12^*(k^*2^\wedge 7+106)+9$ do not have the images at the RHT

Let's continue creating columns with numbers 4 and further to infinity

The resulting structure is called the **Reverse Structure of the Collatz**

It is obvious that in the Reverse Structure of the Collatz

each column, starting from the 1st, contains $2^{(n-1)}$ new sequences of horizontal natural integers of type 3, 5, 7, 9, 11 and 1 (n column number), which in the previous columns did not occur

We define another new structure, which we call the **Reverse Fibonacci Structure of the Collatz**

In the column with the number 1 of the **Reverse Fibonacci Structure** of the Collatz we arrange the sequence of elements

$12*(k*2^2+0)+ 3$
$12*(k*2^2+1)+ 7$
$12*(k*2^2+2)+11$

In the column with the number 2 of the **Reverse Fibonacci Structure** of the Collatz we arrange the sequence of elements

$12*(k*2^3+7)+ 3$
$12*(k*2^3+4)+ 7$
$12*(k*2^3+1)+11$

In the column with the number 3 of the **Reverse Fibonacci Structure** of the Collatz we arrange the sequence of elements

$12*(k*2^4+6)+ 3$
$12*(k*2^4+1)+ 3$
$12*(k*2^4+11)+ 7$
$12*(k*2^4+6)+ 7$
$12*(k*2^4+0)+11$
$12*(k*2^4+11)+11$

In the column with the number 4 of the **Reverse Fibonacci Structure** of the Collatz we arrange the sequence of elements

$12*(k*2^5+26)+ 3$
$12*(k*2^5+11)+ 3$
$12*(k*2^5+29)+ 3$

12*(k*2^5+15)+ 7
12*(k*2^5+0)+ 7
12*(k*2^5+18)+ 7
12*(k*2^5+4)+11
12*(k*2^5+21)+11
12*(k*2^5+7)+11

In the column with the number 5 of the **Reverse Fibonacci Structure** of the Collatz we arrange the sequence of elements

12*(k*2^6+46)+ 3
12*(k*2^6+3)+ 3
12*(k*2^6+18)+ 3
12*(k*2^6+57)+ 3
12*(k*2^6+5)+ 3
12*(k*2^6+3)+ 7
12*(k*2^6+24)+ 7
12*(k*2^6+39)+ 7
12*(k*2^6+14)+ 7
12*(k*2^6+26)+ 7
12*(k*2^6+24)+11
12*(k*2^6+45)+11
12*(k*2^6+60)+11
12*(k*2^6+35)+11
12*(k*2^6+47)+11

In the column with the number 0 of the **Reverse Fibonacci Structure of the Collatz** we arrange the sequence of elements

12*(2*k+0)+5
12*(2*k+1)+9
12*(2*k+1)+1

In the column with the number 2 of the **Reverse Fibonacci Structure of the Collatz** we arrange too the sequence of elements

12*(k*2^3+4)+ 1

178

12*(k*2^3+1)+ 5
12*(k*2^3+6)+ 9

In the column with the number 3 of the **Reverse Fibonacci Structure of the Collatz** we arrange the sequence of elements too

12*(k*2^4+2)+ 1
12*(k*2^4+7)+ 5
12*(k*2^4+12)+ 9

In the column with the number 4 of the **Reverse Fibonacci Structure of the Collatz** we arrange too the sequence of elements

12*(k*2^5+16)+ 1
12*(k*2^5+6)+ 1
12*(k*2^5+5)+ 5
12*(k*2^5+27)+ 5
12*(k*2^5+26)+ 9
12*(k*2^5+16)+ 9

In the column with the number 5 of the **Reverse Fibonacci Structure of the Collatz** we arrange too the sequence of elements

12*(k*2^6+24)+ 1
12*(k*2^6+58)+ 1
12*(k*2^6+30)+ 1
12*(k*2^6+45)+ 5
12*(k*2^6+15)+ 5
12*(k*2^6+51)+ 5
12*(k*2^6+2)+ 9
12*(k*2^6+36)+ 9
12*(k*2^6+8)+ 9

In the column with the number 6 of the **Reverse Fibonacci Structure of the Collatz** we arrange too the sequence of elements

12*(k*2^7+64)+ 1
12*(k*2^7+106)+ 1
12*(k*2^7+8)+ 1
12*(k*2^7+86)+ 1

12*(k*2^7+110)+ 1
12*(k*2^7+21)+ 5
12*(k*2^7+63)+ 5
12*(k*2^7+93)+ 5
12*(k*2^7+43)+ 5
12*(k*2^7+67)+ 5
12*(k*2^7+106)+ 9
12*(k*2^7+20)+ 9
12*(k*2^7+50)+ 9
12*(k*2^7+0)+ 9
12*(k*2^7+24)+ 9

Denote by F[n] Fibonacci numbers F[0] = 1, F[1] = 1, F[2]= 2, F[3] = 3, F[4] = 5,........

From the above we see that the number of sequences (natural integers of type 3,7,11) in each of the first 5 columns (starting with 1) of the **Reverse Fibonacci Structure of the Collatz** is the Fibonacci sequence

From the above we also see that the number of sequences (natural integers of type 5,9,1) in each of the first 5 columns (starting with 2) of the **Reverse Fibonacci Structure of the Collatz** is the Fibonacci sequence

We will show that the numbers of sequences in each column of the **Reverse Fibonacci Structure of the Collatz** is the Fibonacci sequence

Let the column with the number m-1 have F1 sequences of numbers 12*(k*2^m+a1[i])+ 1 (where m>5 and i=0,.., F1-1)

From the previous sections we know that one part of such sequences consists of images of the reverse horizontal transformation of every third natural integer of type 1, in which the factor $2^{(m-2)}$ is present.

From the previous sections we know that the other part of such sequences consists of images of the reverse horizontal transformation of every third natural integer of type 7, in which the factor $2^{(m-2)}$ is present.

Therefore
if the number of sequences of reverse horizontal transformation of natural integers of type 1 in which the factor $2^{(m-2)}$ is present is F[m-2]
and
if the number of sequences of reverse horizontal transformation of natural integers of type 7 in which the factor $2^{(m-2)}$ is present is F[m-1]
then F1 = F[m]

Let the column with the number m-1 have F5 sequences of natural integers $12*(k*2^m+a5[i])+5$ (where m>5 and i=0,.., F5-1)

From the previous sections we know that one part of such sequences consists of images of the reverse horizontal transformation of every third natural integer of type 1, in which the factor $2^{(m-2)}$ is present.

From the previous sections we know that the other part of such sequences consists of images of the reverse horizontal transformation of every third natural integer of type 7, in which the factor $2^{(m-2)}$ is present.

Therefore
if the number of sequences of reverse horizontal transformation of natural integers of type 1 in which the factor $2^{(m-2)}$ is present is F[m-2]
and

if the number of sequences of reverse horizontal transformation of natural integers of type 7 in which the factor $2^{\wedge}(m-2)$ is present is $F[m-1]$
then $F5 = F[m]$

Let the column with the number m-1 have F9 sequences of natural integers $12*(k*2^{\wedge}m+a9[i])+ 9$ (where m>5 and i=0,.., F9-1)

From the previous sections we know that one part of such sequences consists of images of the reverse horizontal transformation of every third natural integer of type 1, in which the factor $2^{\wedge}(m-2)$ is present.

From the previous sections we know that the other part of such sequences consists of images of the reverse horizontal transformation of every third natural integer of type 7, in which the factor $2^{\wedge}(m-2)$ is present.

Therefore
if the number of sequences of reverse horizontal transformation of natural integers of type 1 in which the factor $2^{\wedge}(m-2)$ is present is $F[m-2]$
and
if the number of sequences of reverse horizontal transformation of natural integers of type 7 in which the factor $2^{\wedge}(m-2)$ is present is $F[m-1]$
then $F9 = F[m]$

Let the column with the number m-1 have F3 sequences of natural integers $12*(k*2^{\wedge}m+a3[i])+ 3$ (where m>5 and i=0,.., F3-1)

From the previous sections we know that one part of such

sequences consists of images of the reverse horizontal transformation of every third natural integer of type 5, in which the factor $2^{(m-1)}$ is present.

From the previous sections we know that the other part of such sequences consists of images of the reverse horizontal transformation of every third natural integer of type 11, in which the factor $2^{(m-1)}$ is present.

Therefore
if the number of sequences of reverse horizontal transformation of natural integers of type 5 in which the factor $2^{(m-1)}$ is present is F[m-2]
and
if the number of sequences of reverse horizontal transformation of natural integers of type 11 in which the factor $2^{(m-1)}$ is present is F[m-1]
then F3 = F[m]

Let the column with the number m-1 have F7 sequences of natural integers $12*(k*2^m+a7[i])+ 7$ (where m>5 and i=0,.., F7-1)

From the previous sections we know that one part of such sequences consists of images of the reverse horizontal transformation of every third natural integer of type 5, in which the factor $2^{(m-1)}$ is present.

From the previous sections we know that the other part of such sequences consists of images of the reverse horizontal transformation of every third natural integer of type 11, in which the factor $2^{(m-1)}$ is present.

Therefore
if the number of sequences of reverse horizontal transformation of natural integers of type 5 in which the

factor $2^{(m-1)}$ is present is $F[m-2]$
and
if the number of sequences of reverse horizontal
transformation of natural integers of type 11 in which the
factor $2^{(m-1)}$ is present is $F[m-1]$
then $F7 = F[m]$

Let the column with the number m-1 have F11 sequences
of natural integers $12*(k*2^m+al1[i])+ 11$ (where m>5
and i=0,.., F11-1)

From the previous sections we know that one part of such
sequences consists of images of the reverse horizontal
transformation of every third natural integer of type 5, in
which the factor $2^{(m-1)}$ is present.

From the previous sections we know that the other part of
such sequences consists of images of the reverse horizontal
transformation of every third natural integer of type 11, in
which the factor $2^{(m-1)}$ is present.

Therefore
if the number of sequences of reverse horizontal
transformation of natural integers of type 5 in which the
factor $2^{(m-1)}$ is present is $F[m-2]$
and
if the number of sequences of reverse horizontal
transformation of natural integers of type 11 in which the
factor $2^{(m-1)}$ is present is $F[m-1]$
then $F11 = F[m]$

The above shows that the number of sequences of images
in which there are factors $2^1, 2^2, 2^3, 2^3, 2^4$ represent
a series of Fibonacci

Therefore, by induction, the number of sequences of images in which the factor $2^{\wedge}(m)$ is present is a series of Fibonacci

Let $s > 0$ then let us consider what is the number of horizontal natural integers of types 3, 5, 7, 9, 11 and 1, which are less than $12*2^{\wedge}s$ and belong to the Reverse Fibonacci the Structure of Collatz

Column number 1 contains $F[0]*2^{\wedge}(s-1)$ numbers of each type 3, 7, 11 that are less than $12*2^{\wedge}(s+1)$, $(s => 1)$

Column number 1 does not contain numbers of each type 5, 9, and 1

Column number 2 contains $F[1]*2^{\wedge}(s-2)$ numbers of each type 3, 7, 11 that are less than $12*2^{\wedge}(s+1)$, $(s => 2)$

Column number 2 contains $F[0]*2^{\wedge}(s-2)$ numbers of each type 5, 9, 1 that are less than $12*2^{\wedge}(s+1)$, $(s => 2)$

Column number 3 contains $F[2]*2^{\wedge}(s-3)$ numbers of each type 3, 7, 11 that are less than $12*2^{\wedge}(s+1)$, $(s => 3)$

Column number 3 contains $F[1]*2^{\wedge}(s-3)$ numbers of each type 5, 9, 1 that are less than $12*2^{\wedge}(s+1)$, $(s => 3)$

Column number n contains $F[n-1]*2^{\wedge}(s-n)$ numbers of each type 3, 7, 11 that are less than $12*2^{\wedge}(s+1)$, $(s => n)$

Column number n contains $F[n-2]*2^{\wedge}(s-n)$ numbers of each type 5, 9, 1 that are less than $12*2^{\wedge}(s+1)$, $(s => n)$

Let the number Summa3_7_11(n,s) be the sum of the natural integers in the first n (n>0) columns, where each of the natural integers of each type is 3, 7, 11 less than

12*2^(s+1), n < s+1

Summa(n,s)3_7_11/(2^(s+1)) = F[0]*(2^(s-1))/(2^(s+1)) + F[1]*(2^(s-2))/(2^(s+1)) + .. + F[n-1]*(2^(s-n))/(2^(s+1)) = F[0]/(2^(2)) + F[1]/(2^(3)) + .. + F[n-1]/(2^(n+1))

It is obvious that Summa3_7_11(n,s)/(2^(s+1)) <= 1
It is also obvious that
Summa3_7_11(n1,s)/(2^(s+1)) <
Summa3_7_11(n2,s)/(2^(s+1)) если n1 < n2

Hence the sequence Sumo 3_7_11(n,s)/(2^(s+1)) has a limit S when n tends to infinity
Hence the sequenceSumma3_7_11(n,s)/(2^(s+1)) has a limit S when n tends to infinity

S = 1/2^2 + 1/2^3 + 2/2^4 + 3/2^5 + 5/2^6 + 8/2^7 + ...

(1/2)*S = 1/2^3 + 1/2^4 + 2/2^5 + 3/2^6 + 5/2^7 + ...

subtract the second line from the first line

(1/2)*S = 1/2^2 + 0/2^3 + 1/2^4 + 1/2^5 + 2/2^6 + 3/2^7 + ...

(1/2)*S = 1/2^2 + (1/2^2)*(1/2^2 + 1/2^3 + 2/2^4 + 3/2^5 + 5/2^6 + 8/2^7 + ...)

(1/2)*S = 1/2^2 + (1/2^2)*S

(2^2)(1/2)*S = (2^2)(1/2^2 + (1/2^2)*S)

2*S = 1 + S

S = 1

Let the number Summa5_9_1(n,s) be the sum of the natural integers in the first n (n>0) columns, where each of the natural integers of each type is 5, 9, 1 less than $12*2^{(s+1)}$, n < s+1

It is obvious that Summa5_9_1(n,s) = (1/2)*3_7_11

And hence the sequence Summa5_9_1(n,s)/($2^{(s+1)}$)) has a limit S/2 when n tends to infinity namely ½

At the same time every second natural integer of type 5, 9 and 1 is a vertical natural integer and hence the corresponding sequence which includes all natural integers of type 5, 9 and 1 has a limit S when n tends to infinity namely 1

If in the Reverse Structure of the Collatz and, accordingly, in the Reverse Fibonacci Structure of the Collatz there are all odd natural numbers greater than 1, then all direct horizontal sequences are finite, because in this case from any odd natural number you can get into a vertical natural number of type 5, 9 or 1 for a finite number of steps, consistently applying direct horizontal transformation

Denote by SummaReverse(n, s) the sum of the number of all natural integers in the first n (n>0) columns that are less than $12*2^{(s+1)}$, n < s+1
((namely Summa3_7_11(n, s), Summa5_9_1 (n,s) and the sum of the corresponding vertical numbers)

Suppose that there is some odd natural integer A (A!= 1), which is not in the Inverse structure of the Collatz and, accordingly, in the Reverse Fibonacci structure of the Collatz, then there is a natural integer N such that
A < $12*2^N$ and $12*2^{(N-1)}$ < A

The natural integer A can only be a horizontal natural integer, because the Reverse structure of the Collatz and, accordingly, the Reverse Fibonacci structure of the Collatz by definition contain all vertical natural integers, and we assumed that the natural integer A does not belong to the Reverse Structure of the Collatz and therefore does not belong to the Reverse Fibonacci Structure of the Collatz

The natural integer A cannot be a natural integer of type 3 or type 9 because the natural integers of this type have no images in the reverse horizontal transformation

If for some $k > 0$, the natural integer $A[k] = k*12*2^N + A$ belongs to Reverse the Structure of Collatz and it belongs therefore Reverse Fibonacci the Structure of Collatz, then the natural integers $A[k] = k*12*2^N + A$ belong to Reverse the Structure of Collatz and therefore belong to Reverse Fibonacci the Structure of Collatz is for any $k = 0,1,2,...$

Thus, if the natural integers A does not belong to the Reverse Structure of the Collatz and, accordingly, in the Reverse Fibonacci Structure of the Collatz, then the entire sequence of natural integers $A[k] = k*12*2^{(N-1)} + A$ for any $k = 0,1,2,...$ does not belong to the Reverse structure of the Collatz and, respectively, in the Reverse Fibonacci Structure of the Collatz

If the number A is a natural integer of type 5 or type 11, then all horizontal natural integers in the sequence of natural integers $A[k]$ in this case have images of natural integers of type 3, 7, or 11 in the reverse horizontal transformation and which can be represented as

$$B3[k] = 12*(k*2^N + b3) + 3, b3 < 2^N$$
$$B7[k] = 12*(k*2^N + b7) + 7, b7 < 2^N$$

$$B11[k] = 12*(k*2^N + b11) + 11, b11 < 2^N$$

The natural integers of $B3[k]$ are not images under the reverse horizontal transform

In turn, the natural integers $B7[k]$ and the natural integers $B11[k]$ have images in the reverse horizontal transformation

These images can be represented as

$$C1[k] = 12*(k*2^{(N+2)} + c1) + 1, c1 < 2^{(N+2)}$$
$$C5[k] = 12*(k*2^{(N+2)} + c5) + 5, c5 < 2^{(N+2)}$$
$$C9[k] = 12*(k*2^{(N+2)} + c9) + 9, c9 < 2^{(N+2)}$$
$$C3[k] = 12*(k*2^{(N+1)} + c3) + 3, c3 < 2^{(N+1)}$$
$$C7[k] = 12*(k*2^{(N+1)} + c7) + 7, c7 < 2^{(N+1)}$$
$$C11[k] = 12*(k*2^{(N+1)} + c11) + 11, c11 < 2^{(N+1)}$$

If the natural integer A is a natural integer of type 7 or type 1, then all horizontal natural integers in the sequence of natural integers $A[k]$ in this case have images of natural integers of type 5, 9, or 1 in the reverse horizontal transformation and which can be represented as

$$D5[k] = 12*(k*2^{(N+1)} + d5) + 5, d5 < 2^{(N+1)}$$
$$D9[k] = 12*(k*2^{(N+1)} + d9) + 9, d9 < 2^{(N+1)}$$
$$D1[k] = 12*(k*2^{(N+1)} + d1) + 1, d1 < 2^{(N+1)}$$

The natural integers of $D9[k]$ are not images under the reverse horizontal transform

In turn, the natural integers $D5[k]$ and the natural integers $D1[k]$ have images in the reverse horizontal transformation

These images can be represented as

E1[k] = 12*(k*2^(N+2) +e1) + 1, e1 < 2^(N+2)
E5[k] = 12*(k*2^(N+2) +e5) + 5, e5 < 2^(N+2)
E9[k] = 12*(k*2^(N+2) +e9) + 9, e9 < 2^(N+2)
E3[k] = 12*(k*2^(N+1) +e3) + 3, e3 < 2^(N+1)
E7[k] = 12*(k*2^(N+1) +e7) + 7, e7 < 2^(N+1)
E11[k] = 12*(k*2^(N+1) +e11) + 11, e11 < 2^(N+1)

Denote by F1[k], F7[k], F5[k], F11[k] the natural integers
of sequences C1[k], C7[k], C5[k], C11[k] or the natural
integers of sequences E1[k], E7[k], E5[k], E11[k]

Place the natural integers of sequences F1[k], F7[k], F5[k],
F11[k] for any k = 0,1,2, ... in the column with number 1

In the column with the number 2 we place the images of the
natural integers F1[k], F7[k], F5[k], F11[k] in the reverse
horizontal transformation

Let's continue creating columns to infinity

The resulting structure is called **Pseudo-Reverse Structure
of the Collatz**

We define another new structure, which we call the
Pseudo-Reverse Fibonacci Structure of the Collatz

In the column with the number 1 of the Pseudo-Reverse
Fibonacci Structure of the Collatz, we place the sequences
of elements
F3[k] = 12*(k*2^(N+1) +f3) + 3, f3 < 2^(N+1)
F7[k] = 12*(k*2^(N+1) +f7) + 7, f7 < 2^(N+1)
F11[k] = 12*(k*2^(N+1) +f11) + 11, f11 < 2^(N+1)

In the column with the number 2 of the Pseudo-Reverse
Fibonacci Structure of the Collatz, we place the sequences
of elements that contain the factor 2^(N+2)

In the column with the number 3 of the Pseudo-Reverse Fibonacci Structure of the Collatz, we place the sequences of elements that contain the factor $2^{\wedge}(N+3)$

…..

In the column with the number n of the Pseudo-Reverse Fibonacci Structure of the Collatz, we place the sequences of elements that contain the factor $2^{\wedge}(N+n)$

By analogy with the **Reverse Fibonacci Structure of the Collatz**, the number of sequences in each column of the **Pseudo Reverse Fibonacci Structure of the Collatz** is the Fibonacci sequence

Let the number SummaPseudoReverse(n,s) be the sum of all the natural integers of the first n columns of the Pseudo Inverse Fibonacci structure of the Collatz, where each of the natural integers of each type is 3, 5, 7, 9, 11 and 1, which are less than $12*2^{\wedge}(s)$, $n < s+1$

It is obvious
thatlimit(SummaPseudoReverse(n,s)/($2^{\wedge}(s+N+2)$) > 0

and consequently

limit(SummaReverse(n,s)/($2^{\wedge}s$) +
limit(SummaPseudoReverse(n,s)/($2^{\wedge}(s+N+2)$)) > 1

which is impossible because the total sum of the natural integers from the Reverse Structure of the Collatz and any number of Pseudo-Reverse Structures of the Collatz is always no more than the total number of odd natural numbers for any subset

Thus, our assumption that there may be an odd natural number that is not included in the Inverse structure of the Collatz is incorrect

However, the fact that any odd natural number is included in the Reverse Structure of the Collatz means that any DIRECT horizontal sequence is finite. Namely, a DIRECT horizontal sequence started in any odd natural number A (A != 1) always ends in a vertical number of type 5, 9, or 1

15.2 All Reverse horizontal sequences are finite

15.2.1 The First proof of the finiteness of the reverse horizontal sequences

Suppose that there is an infinite reverse sequence of horizontal natural integers of type 5, 7, 11 or 1
As follows from sections 13.1.1-13.1.3, all natural integers in this sequence are different
Thus, each sequence that starts at any of these natural integers is infinite
Therefore, the "density" of the distribution of sequences of arbitrarily large length is greater than 0
However, in previous sections 13.1.1-13.1.2 it was shown that the "density" of straight horizontal sequences whose length is greater than or equal to n is $(2/3)^n$
That is, the" density " of the distribution of such sequences tends to zero and therefore the assumption of an infinite inverse horizontal sequence is incorrect

15.2.2 The second proof of finiteness of reverse horizontal sequences. Direct Structure of the Collatz

Further everywhere in section 15.2.2 the symbol → will denote the DIRECT horizontal transformation F and s will denote the natural integer
We further show that ALL REVERSE horizontal sequences are of finite length
Dear readers, there is no typo in this place
The following will show that the REVERSE horizontal sequences have a finite length. This prove is based on the properties of the direct horizontal sequences

Consider the following structure, which consists of columns.

Everywhere further k=0,1,2,…

In the column with the number 0 we place natural integers of type 3 and non type 9

12*(2*k+0)+3
12*(2*k+1)+3
12*(4*k+0)+9
12*(4*k+1)+9
12*(4*k+2)+9
12*(4*k+3)+9

the natural integers 12*(4*k+1)+9 and 12*(4*k+3)+9 are vertical and have no images in direct horizontal transformation

all other natural integers have images with direct horizontal transformation let's place these images in the column with the number 1

$12^*(2^*k+0)+3 \rightarrow 12^*(k^*3\wedge1+0) +5,$
$12^*(2^*k+1)+3 \rightarrow 12^*(k^*3\wedge1+1) +11$
$12^*(4^*k+0)+9 \rightarrow 12^*(k^*3\wedge1+0) +7$
$12^*(4^*k+2)+9 \rightarrow 12^*(k^*3\wedge1+2) +1$

natural integers $12^*((4^*k+0)^*3\wedge1+0) +5,$
$12^*((4^*k+2)^*3\wedge1+0) +5, 12^*(k(4^*k+1)^*3\wedge1+2) +1,$
$12^*(k(4^*k+3)^*3\wedge1+2) +1$ are vertical and have no images
in direct horizontal transformation

all other natural integers have images with direct horizontal
transformation

let's place these images in the column with the number 2

$12^*((4^*k+0)^*3\wedge1+0) +5 =$ vertical number
$12^*((4^*k+1)^*3\wedge1+0) +5 \rightarrow 12^*(k^*3\wedge2+2) +7$
$12^*((4^*k+2)^*3\wedge1+0) +5 =$ vertical number
$12^*((4^*k+3)^*3\wedge1+0) +5 \rightarrow 12^*(k^*3\wedge2+7) +1$

$12^*((2^*k+0)^*3\wedge1+1) +11 \rightarrow 12^*(k^*3\wedge2+2) +11$
$12^*((2^*k+1)^*3\wedge1+1) +11 \rightarrow 12^*(k^*3\wedge2+7) +5$

$12^*((2^*k+0)^*3\wedge1+0) +7 \rightarrow 12^*(k^*3\wedge2+0) +11$
$12^*((2^*k+1)^*3\wedge1+0) +7 \rightarrow 12^*(k^*3\wedge2+5) +5$

$12^*(k(4^*k+0)^*3\wedge1+2) +1 \rightarrow 12^*(k^*3\wedge2+1) +7$
$12^*(k(4^*k+1)^*3\wedge1+2) +1 =$ vertical number
$12^*(k(4^*k+2)^*3\wedge1+2) +1 \rightarrow 12^*(k^*3\wedge2+6) +1$
$12^*(k(4^*k+3)^*3\wedge1+2) +1 =$ vertical number

194

Let's continue creating columns with numbers 3 and further to infinity

The resulting structure is called the **Direct Structure of the Collatz**

It is obvious that in the Direct Structure of the Collatz each column, starting from the 1st, contains $2^{(n-1)}$ new sequences of horizontal natural integers of type 5, 7, 11 and 1 (n column number), which in the previous columns did not occur

Let $s > 0$ then consider how many natural integers of type 5, 7, 11 and 1 that are less than $12*3^s$ and belong to the Direct Structure of the Collatz

Column number 1 contains $(2^{(0)})*(3^{(s-1)})$ natural integers of each type 5, 7, 11 and 1 that are less than $12*3^s$, $(s => 1)$

Column number 2 contains $(2^{(1)})*(3^{(s-2)})$ natural integers of each type 5, 7, 11 and 1 that are less than $12*3^s$, $(s => 2)$

Column number 3 contains $(2^{(2)})*(3^{(s-3)})$ natural integers of each type 5, 7, 11 and 1 that are less than $12*3^s$, $(s => 3)$

Column number n contains $(2^{(n-1)})*(3^{(s-n)})$ natural integers of each type 5, 7, 11 and 1 that are less than $12*3^s$, $(s => n)$

Let the number Summa (n,s) be the sum of all the natural integers of the first n columns, where each of the numbers of each type is 5, 7, 11 and 1, which are less than $12*3^s$, $n < s+1$

Then the sequence Summa(n,s)/(3^s) tends to 1 when n tends to infinity

Summa(n,s)/(3^s) = (2^(0))*(3^(s-1))/(3^s) + (2^(1))*(3^(s-2))/(3^s) + .. + (2^(n-1))*(3^(s-n))/(3^s) = (2^(0))/(3^(1)) + (2^(1))/(3^(2)) + .. + (2^(n-1))/(3^(n)) = {{1 - (2^1)/(3^1)}*{(2^0)/(3^0) + (2^1)/(3^1) + .. + (2^(n-1))/(3^(n-1))}}/(3^(1)) = {1 – (2^n)/(3^n)}/{1 - (2^1)/(3^1)}*(3^(1)) = {1 – (2^n)/(3^n)}

If the Direct structure of the Collatz contains all odd natural numbers greater than 1, then all inverse horizontal sequences are finite, because in this case from any odd natural number you can get to the number of type 3 or 9 in a finite number of steps, consistently applying the inverse horizontal transformation

Suppose that there is some odd natural integer A (A!= 1), which is not in the Direct structure of the Collatz, then there is a number N such that
A < 12*3^N and 12*3^(N-1) < A

The number A can only be of type 5,7,11 or 1, because the Direct Structure of the Collatz by definition contains all the numbers of type 3 and 9, and we assumed that the natural integer A does not belong to the Direct Structure of the Collatz

Suppose that there is some odd natural integer A (A!= 1), which is not in Direct Structure of the Collatz then there is a natural integer N such that
A < 12*3^N and 12*3^(N-1) < A

The natural integer A can only be of type 5,7,11 or 1, because the Direct Structure of the Collatz by definition contains all the numbers of type 3 and 9, and we assumed that the natural integer A does not belong to the Direct Structure of the Collatz

If for some $k > 0$, the natural integer $A[k] = k*12*3^N + A$ belongs to Direct the Structure of Collatz, then the natural integers $A[k] = k*12*3^N + A$ belong to Direct Structure of Collatz for any $k = 0,1,2,\ldots$

Thus, if the natural integers A does not belong to the Direct Structure of the Collatz, then the entire sequence of natural integers $A[k] = k*12*3^N + A$ for any $k = 0,1,2,\ldots$ does not belong to the Direct Structure of the Collatz

If the natural integer A is a natural integer of type 5 or 1, then every second natural integer in the sequence $A[k]$ is vertical and has no image in the direct horizontal transformation

All horizontal natural integers in the sequence of natural integers $A[k]$ in this case have images of natural integers of type 1 and 7, which can be represented as

$B1[k] = 12*(k*3^{(N+1)} + b1) + 1, b1 < 3^{(N+1)}$
$B7[k] = 12*(k*3^{(N+1)} + b7) + 7, b7 < 3^{(N+1)}$

In turn, half of the natural integers $B1[k]$ and all the natural integers $B7[k]$ are horizontal natural integers and have images in direct horizontal transformation

These images can be represented as

$C1[k] = 12*(k*3^{(N+2)} + c1) + 1, c1 < 3^{(N+2)}$
$C7[k] = 12*(k*3^{(N+2)} + c7) + 7, c7 < 3^{(N+2)}$
$C5[k] = 12*(k*3^{(N+2)} + c5) + 5, c5 < 3^{(N+2)}$
$C11[k] = 12*(k*3^{(N+2)} + c11) + 11, c11 < 3^{(N+2)}$

If the natural integer A is a natural integer of type 7 or 11, then all the natural integers in the sequence A[k] are horizontal and have an image with a direct horizontal transformation

Horizontal natural integers in the sequence of natural integers A[k] in this case have images of natural integers of type 5 and 11, which can be represented as

$D5[k] = 12*(k*3^{(N+1)} +d5) + 5, d5 < 3^{(N+1)}$
$D11[k] = 12*(k*3^{(N+1)} +d11) + 11, d11 < 3^{(N+1)}$

In turn, half of the natural integers D5[k] and all the natural integers D11[k] are horizontal natural integers and have images in direct horizontal transformation

These images can be represented as

$E1[k] = 12*(k*3^{(N+2)} +e1) + 1, e1 < 3^{(N+2)}$
$E7[k] = 12*(k*3^{(N+2)} +e7) + 7, e7 < 3^{(N+2)}$
$E5[k] = 12*(k*3^{(N+2)} +e5) + 5, e5 < 3^{(N+2)}$
$E11[k] = 12*(k*3^{(N+2)} +e11) + 11, e11 < 3^{(N+2)}$

Denote by F1[k], F7[k], F5[k], F11[k] the numbers of sequences C1[k], C7[k], C5[k], C11[k] or the numbers of sequences E1[k], E7[k], E5[k], E11[k]

Let us to place the natural integers of sequences F1[k], F7[k], F5[k], F11[k] for any k = 0,1,2, ... in the column with number 1

In the column with number 2 we place images of natural integers F1[k], F7[k], F5[k], F11[k] with direct horizontal transformation

Let's continue creating columns to infinity

The resulting structure is called **Pseudo-Direct structure of the Collatz**

Let SummaPseudo (n, s) be the sum of all the natural integers of the first n columns of the Pseudo-Straight structure of the Collatz, where each of the natural integers of each type is 5, 7, 11 and 1, which are less than $12*3^{\wedge}(s)$, $n < s+1$

Then the sequence SummaPseudo(n,s)/($3^{\wedge}(s+N+2)$) tends to $1/(3^{\wedge}(N+2))$ when n tends to infinity

SummaPseudo(n,s)/($3^{\wedge}(s+N+2)$) = $(2^{\wedge}(0))*(3^{\wedge}(s-1))/(3^{\wedge}s(s+N+2)) + (2^{\wedge}(1))*(3^{\wedge}(s-2))/(3^{\wedge}(s+N+2)) + .. + (2^{\wedge}(n-1))*(3^{\wedge}(s-n))/(3^{\wedge}(s+N+2)) = (2^{\wedge}(0))/(3^{\wedge}(1)) + (2^{\wedge}(1))/(3^{\wedge}(2)) + .. + (2^{\wedge}(n-1))/(3^{\wedge}(n))/(3^{\wedge}(N+2)) = \{(2^{\wedge}0)/(3^{\wedge}0) + (2^{\wedge}1)/(3^{\wedge}1) + .. + (2^{\wedge}(n-1))/(3^{\wedge}(n-1))\}/(3^{\wedge}(1)) = \{1 - (2^{\wedge}1)/(3^{\wedge}1)\}*\{1 – (2^{\wedge}n)/(3^{\wedge}n)\}/(3^{\wedge}(1)) = \{1 – (2^{\wedge}n)/(3^{\wedge}n)\}/(3^{\wedge}(N+2))$

Thus, for n tending to infinity

limit(Summa(n,s)/($3^{\wedge}s$) + limit(SummaPseudo(n,s)/($3^{\wedge}(s+N+2)$)) > 1

which is impossible because the total sum of the natural integers from a Direct Structure of Collatz and any number of Pseudo Direct Structures of Collatz is always no more than the total natural integers of odd positive integers for any subset

Thus, our assumption that there may be an odd natural number that is not included in the Direct structure of the Collatz is incorrect

199

**However, the fact that any odd natural integer is
included in the Direct structure of the Collatz means
that any reverse horizontal sequence is finite. Namely,
the reverse horizontal sequence started in any odd
natural integer A (a != 1) will always end in the natural
integer of type 3 or type 9**

15.3 complete horizontal and complete vertical sequences

In the previous sections we established that any direct and
any reverse horizontal sequence is finite.

Thus, any odd natural integer A (a != 1) always belongs to
some finite horizontal sequence with a natural integer of
type 3 or 9 at one end and a vertical natural integer of type
5, 9 or 1 at the other.

We call such a horizontal sequence the **Complete
Horizontal Sequence**

Note: If the number A is a vertical natural integer of type 9,
then the Complete Horizontal Sequence consists only of
this number. In all other cases, the length of this sequence
is greater than 1.

We call a **Complete Vertical Sequence** a sequence of
natural odd natural integers that starts at any horizontal
natural integer or root natural integer 1 and continues with
the Complete Trunk that is generated by that horizontal or
root natural integer

16. The Uniqueness of The Canonical Tree of Collatz

Denote by **Vertical_1** the set of all complete vertical sequences
Denote by **Vertical_2** also the set of all complete vertical sequences
Denote by **Horizontal_1** the set of all complete horizontal sequences
Denote by **Horizontal_2** also the set of all complete horizontal sequences

We construct two new sets V and G

The set V will be constructed as follows

We take another complete vertical sequence from the set **Vertical_1** and we connect the corresponding complete horizontal sequence from the set **Gorizontal_1** to each vertex of this complete vertical sequence

It is obvious that the necessary complete horizontal sequence is always there

It is also obvious that when we do this with all complete vertical sequences from the set **Vertical_1**, there will be no elements left in the set **Gorizontal_1**

The set G will be constructed as follows

we'll take the next full horizontal sequence of Gorizontal_2 and to each vertex of this sequence, we connect the corresponding complete vertical sequence of the plurality of Vertical_2

It is obvious that the necessary complete vertical sequence

is always there

It is also obvious that when we do this with all complete horizontal sequences from the **Gorizontal_2** set, there will be no elements left in the **Vertical_2** set

Next, we do any of the two procedures

The first procedure is as follows.
We take another element from the set V and glue it with all elements of the set G which have the same complete horizontal sequences as in the this element from the set V

It is obvious that when we do this with all elements of the set V, there will be no elements in the set G

The second procedure is as follows.
We take another element from the set G and glue it with all elements of the set V which have the same complete vertical sequences as in the another element from the set G

It is obvious that when we do this with all elements of the set G, there will be no elements in the set V

As a result of any of the above two procedures, we get some trees

In any case, among these trees will necessarily be the Canonical tree (that is, the tree "grown" from the root natural integer 1)

If the result of any of the procedures is obtained only Canonical tree that means that the **Collatz conjecture** is true

All odd natural numbers will be present in the resulting tree From each vertex of the resulting tree moving to the left and down we will always get to the root natural number 1 in a finite number of steps

We will show that the result of any of the above two procedures will always get only one tree, namely the Canonical tree of Collatz.

Suppose that in addition to the Canonical tree of Collatz after the execution of any of above described procedures obtained some more tree

We will take an arbitrary vertex in this tree and start moving left and down the branches and trunks

Our movement will be infinite because we will never be able to get to the root natural integer 1 because it is, in our assumption, in another tree

Thus, we obtain infinitely many vertices (and hence odd natural integers) in which infinite sequences will begin

However, in section 15.1.1 has shown that this is not possible

Therefore, as a result of gluing by any of the two above procedures will always get only one Canonical Tree of Collatz

But the sets Vertical_1, Vertical_2, Gorizontal_1, Gorizontal_2 contain all odd natural numbers and hence the Canonical tree also contains all odd natural integers

But by the definition of a Canonical tree, any vertex of that tree can be reached by moving up and to the right, which is equivalent to the inverse of the vertical and reverse of the horizontal transformation of the Collatz

The reverse transforms of Collatz are in turn one-to-one to direct transforms of Collatz

Therefore, from any vertex of the Canonical tree of Collatz you can get to the root vertex (it means root natural integer 1) in a finite number of steps to the left and down

This movement is equivalent to applying direct horizontal and direct vertical transformations

Therefore, applying such transformations to any odd natural integer, we always get to the root natural integer 1 in a finite number of steps

Thus, any modified reduced sequence that started in any odd natural number not equal to 1 will always reach to the natural integer 1 in a finite number of steps

And hence, by definition of a modified reduced sequence, any Syracuse sequence will always reach to the natural integer 1 in a finite number of steps

17. Afterword

I am appreciated all readers who took the time and energy to read the previous 16 paragraphs.

I beg you to forgive me for the terrible English and a confusing presentation of the material

I assure you these two shortcomings of the book are not a consequence of disrespect for readers

Unfortunately I am in a very difficult financial situation
I'm 70 years old, I get a meager pension of $ 952 and work
three days (no longer allowed) in the Alzheimer's and
Parkinson's ward. I get another $ 325 for this job every two
weeks. My rent is $ 1,400 for a small Studio. To all the
problems I had severe diabetes and all other problems of
my age.

If not for the help of friends and acquaintances I would
have long since died

All my life I have been solving problems that other experts
have said that the solution to these problems is impossible.

The proof of the Collatz Conjecture is one such problem
I wrote in the Preface that the actual evidence can be
criticized and discuss the credibility of some of its
fragments

Paul Erdős 100 years ago said about the Collatz conjecture:
"Mathematics may not be ready for such problems."

What prevented Erdös from moving from the usual
Syracuse sequence to the Reduced one, and from the
reduced to the Modified Reduced sequence, is unclear

What prevented him from Erdös to describe slender
Canonical tree of Collatz is unclear. After these few steps,
it becomes absolutely clear to any student after the age of
12 that the natural odd numbers are arranged in a strict
system and that any of the above sequences ALWAYS
comes in the number 1

For example, in direct conversion of collatz,

to the natural integer of type 1 you can always reach from the "top only and go further down only»,
to the natural integer of type 3 can never get "outside and you can go only up»,
the natural integer of type 5 can be reached " from the down only and go to down only»,
the natural integer of type 7 always possible to get only the "from the top and go up»,
for the type 9 you can never get "outside and you can go only down»,
in the natural integer of type 11 always can be enter only " from below and withdraw only upward."
That is, the natural integer of type 1 is the opposites of the natural integers of type 11 for the natural integer of type 3 are the antipodes of the type 9, the natural integer of type 5 are the antipodes of the type 7.

There are other very interesting patterns
Conclusive proof of the hypothesis requires more effort but is also capable of any student.
What would you do if you were me? I decide describe these simple considerations at the level of English I speak and at the level of presentation I am able to offer?
I chose to write it as best I could in my desperate situation. Because to hide the material described in the book I consider a crime

In turn I am asking everyone whom this material seem interesting please help me materially in any acceptable for you size through my account georgiytyshko@yahoo.com at PayPal

As I wrote in the Preface I am ready to answer any questions and comment on any comments on the live journal page https://howwewanttolive.livejournal.com/

Georgiy Tyshko   March 2019